AA001127

MATERIALS RESEARCH SOCIETY
SYMPOSIUM PROCEEDINGS VOLUME **1737**

# Organic Photovoltaics – Fundamentals, Materials and Devices

November 30 – December 5, 2014
Boston, Massachusetts, USA

**Printed from e-media with permission by:**

Curran Associates, Inc.
57 Morehouse Lane
Red Hook, NY 12571
www.proceedings.com

**ISBN: 978-1-5108-0617-7**

**Some format issues inherent in the e-media version may also appear in this print version.**

©Materials Research Society 2015

This reprint is produced with the permission of the Materials Research Society and Cambridge University Press.

This publication is in copyright, subject to statutory exception and to the provisions of relevant collective licensing agreements. No reproduction of any part may take place without the written permission of Cambridge University Press.

Cambridge University Press
Cambridge, New York, Melbourne, Madrid, Cape Town,
Singapore, São Paulo, Delhi, Tokyo, Mexico City

Cambridge University Press
32 Avenue of the Americas, New York, NY 10013-2473, USA
www.cambridge.org

Materials Research Society
506 Keystone Drive, Warrendale, PA 15086
www.mrs.org

CODEN: MRSPDH

ISBN: 978-1-5108-0617-7

Cambridge University Press has no responsibility for the persistence or accuracy of URLs for external or third-part Internet Web sites referred to in this publication and does not guarantee that any content on such Web sites is, or will remain, accurate or appropriate.

**Additional copies of this publication are available from:**

Curran Associates, Inc.
57 Morehouse Lane
Red Hook, NY 12571 USA
Phone: 845-758-0400
Fax:     845-758-2634
Email:  curran@proceedings.com
Web:    www.proceedings.com

# Organic Photovoltaics - Fundamentals, Materials and Devices

Materials Research Society Symposium Proceedings
Volume 1737

Boston, Massachusetts, USA
30 November - 5 December 2014

# TABLE OF CONTENTS

**Functionalized Rosette Nanotubes as Novel Electron Donor Materials for Solution-Processed Organic Photovoltaics** ........................................................... 1
*L. Shuai, V. Parthasarathy, J.-Y. Cho, T. Yamazaki, R. L. Beingessner, H. Fenniri*

**Computational Modeling of Nanosecond Time-Scale Charge Carrier Dynamics in Organic Semiconductors** ........................................................... 7
*B. Johnson, K. Paudel, O. Ostroverkhova*

**The Influence of Morphology on the Charge Transport in Two-Phase Disordered Organic Systems** ........................................................... 13
*C. F. Woellner, L. D. Machado, P. A. S. Autreto, J. A. Freire, D. S. Galvao*

**Branched Thiophene Oligomer/Polymer Bulk Heterojunction Organic Solar Cell** ........................................................... 19
*F. Martinez, G. Neculqueo, S. O. Vasquez, H. Lemmetyinen, A. Efimov, P. Vivo*

**Organic Photovoltaic Module Development with Inverted Device Structure** ................ 26
*S. Mori, H. Oh-oka, H. Nakao, T. Gotanda, Y. Nakano, H. Jung, A. Iida, R. Hayase, N. Shida, M. Saito, K. Todori, T. Asakura, A. Matsui, M. Hosoya*

**Effective Mg:Ag / MoO$_3$ Recombination Zone for Tandem Organic Photovoltaic Devices** ........................................................... 32
*A. R. Jeong, S. Wiesner, S. Fengler, M. Ch. Lux-Steiner, M. Rusu*

**Author Index**

**Mater. Res. Soc. Symp. Proc. Vol. 1737 © 2015 Materials Research Society**
**DOI: 10.1557/opl.2015.105**

## Functionalized Rosette Nanotubes as Novel Electron Donor Materials for Solution-Processed Organic Photovoltaics

Liang Shuai,[1,2] Venkatakrishnan Parthasarathy,[1,2] Jae-Young Cho,[2] Takeshi Yamazaki,[1,2] Rachel L. Beingessner,[2] Hicham Fenniri[3*]

[1]Department of Chemistry, University of Alberta, 11227 Saskatchewan Drive, Edmonton, Alberta T6G 2G2, Canada [2]National Institute for Nanotechnology, 11421 Saskatchewan Drive, Edmonton, Alberta T6G 2M9, Canada [3]Department of Chemical Engineering, 313 Snell Engineering Center, Northeastern University, 360 Huntington Avenue, Boston, MA 02115-5000

### ABSTRACT

Two self-assembling twin guanine-cytosine (G∧C) hybrid molecules featuring porphyrin (TPPO-(G∧C)$_2$) and oligothiophene groups (6T-(G∧C)$_2$) were synthesized. In organic solution, these molecules self-assemble into one-dimensional rosette nanotubes (RNTs) featuring the porphyrin or oligiothiophene groups on the outer surface. Using a combination of imaging and spectroscopic techniques we established the structure of the TPPO-(G∧C)$_2$ and 6T-(G∧C)$_2$ RNTs and compared the HOMO and LUMO energy levels with PC$_{61}$BM, a well-known electron acceptor material. These studies, in combination with solid-state photoluminescence data of PC$_{61}$BM-TPPO-(G∧C)$_2$ RNT blended thin films, indicates that these self-assembled nanomaterials have great potential as electron donor materials for solution-processed organic photovoltaics.

### INTRODUCTION

Organic photovoltaics (OPVs) have attracted a great deal of interest due to their low cost, light-weight, flexibility and tunability [1]. Solution-processed OPVs are also suitable for large-area device fabrication using roll-to-roll technologies [2]. Intensive studies on the performance of OPV devices using the polymer electron donor (D) material P3HT and the electron acceptor (A) material PC$_{61}$BM, have revealed that the microscopic network structure of the D–A bulk heterojunctions (BHJ) improves the power conversion efficiency (PCE) [3]. The well-defined nanocrystalline morphology of P3HT also significantly enhances the hole mobility and promotes charge transfer and transport. Although thermal annealing is a useful method to impose microscopic ordering on the active layer of the blended D–A thin film, it is not preferred for the mass fabrication of devices. Supramolecular self-assembly alternatively, can provide a convenient pathway to generate highly ordered nanomaterials in all dimensions [4,5]. Although this strategy has been used to improve OPV device performance, only a few examples have been reported to-date [6].

Over the last decade, we have been developing a class of organic nanotubes which are formed through the self-assembly of a G∧C hybrid motif that displays self-complementary hydrogen bonding sites (Figure 1). The twin-G∧C motif (Figure 2) features two G∧C motifs that are covalently linked through an alkyl amine spacer. In solution, single and twin-G∧C molecules self-organize into hexameric super-macrocycles, called rosettes, through the formation of 18- or 36-intermolecular hydrogen bonds, respectively [7,8]. Upon stacking, these rosettes then lead to the formation of linear stacks called rosette nanotubes (RNTs) with an inner channel diameter of *ca.*1.1 nm (Figure 1). Due to the well-defined structure and tolerance for a range of surface functional groups through covalent modification of the G∧C motif [7-10], these nanomaterials have demonstrated great promise in the areas of nanomedicine [11] and catalysis [10].

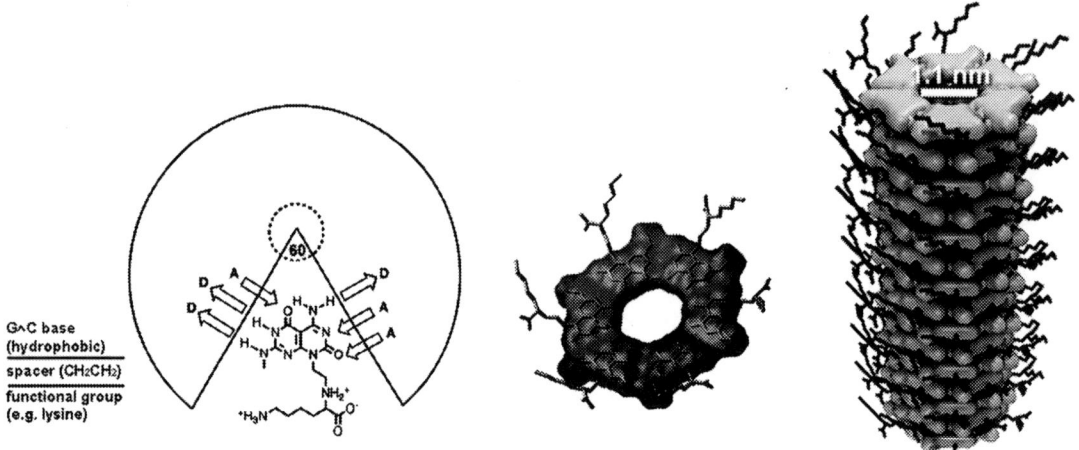

**Figure 1.** Schematic illustration of the hierarchical self-assembly of functionalized G∧C motifs (left) into rosettes (middle) and then RNTs (right).

**Figure 2.** Molecular structures of TPPO-(G∧C)$_2$ and 6T-(G∧C)$_2$.

Herein, we investigate the potential of the RNTs as electron donor materials for solution-processed OPVs. Specifically, we describe the self-assembly and characterization of the twin-G∧C motifs, TPPO-(G∧C)$_2$ and 6T-(G∧C)$_2$ (Figure 2) into RNTs which express porphyrin and hexathiophene units on their outer surface, respectively.

**EXPERIMENT**

As shown in Figure 2, TPPO-(G∧C)$_2$ is the HCl salt of a porphyrin-functionalized twin G∧C building block, while 6T-(G∧C)$_2$ is composed of a hexathiophene unit and a twin G∧C module. The synthesis of the two molecules will be reported elsewhere. All the molecules were fully characterized by $^1$H and $^{13}$C NMR, HRMS and elemental analysis.

**SEM Imaging.** SEM samples were prepared by casting a droplet (5 μL) of the sample solution on a carbon-coated 400-mesh copper grid (Electron Microscopy Sciences). After 10 s the grid was blotted by filter paper, dried in air and then placed under vacuum for 1 d to remove any residual solvent before imaging. SEM imaging was performed on a Hitachi S-4800 field emission high-resolution scanning electron microscope using 15–30 kV accelerating voltage and 15–20 μA current.

**TEM Imaging.** TEM samples were prepared in the same way as the SEM samples, but were also negatively stained using uranyl acetate (0.2% in acetone). TEM imaging was performed on a JEOL 2200FS TEM (200 kV Schottky field emission) with an in-column omega filter and cryo pole piece. Bright field TEM images were acquired using energy filtered zero loss beam (slit width 8 eV). The cross-sectional diameter of a single RNT was determined by randomly measuring individual assemblies using Digital Micrograph software (version 3.9.3 by Gatan). The data is presented in the form of average diameter ± standard deviation (based on 25 measurements).

**Ultraviolet photoelectron spectroscopy (UPS).** The samples were prepared by drop-casting the solution on a freshly etched Si (100) wafer (dipped in 5% HF for 2 min) using a glass pipette. The wafer was dried in a vacuum chamber for 30 min and then transferred into the specimen chamber of the UPS instrument where it remained for 30 min to remove residual solvent before analysis. Room temperature UPS experiments were performed using a Kratos Axis spectrometer with monochromatized Helium I radiation (hv = 21.23 eV) and a hemispherical electron energy analyzer (Kratos Ultra Spectrometer). The system was maintained under ultrahigh vacuum ($< 5 \times 10^{-10}$ Torr), and the power for UPS was 3 kV × 20 mA (60 W). All of the samples were biased at −10 V during the measurements to observe the peak edge of the secondary electron.

## DISCUSSION

TPPO-(G∧C)$_2$ was found to have good solubility (> 20 mg/mL) in a 9:1 (v:v) mixture of 1,2-DCB:MeOH. When a 1.0 mg/mL stock solution of TPPO-(G∧C)$_2$ was heated to the boiling point for 1 min, cooled to room temperature and aged for 24 h, well-dispersed micrometer-long RNTs were observed by SEM imaging (Figure 3A, 0.01 mg/mL). TEM imaging further revealed that the cross-sectional diameter of the RNTs was 5.0 ± 0.3 nm (Figure 3B). In contrast, a stock solution of 6T-(G∧C)$_2$ (1.0 mg/mL) was prepared in 1,2-DCB alone, heated to 120 °C and cooled to room temperature. After aging for 24 h, comparatively shorter nanotubes were observed by SEM imaging (Figure 3C, 0.01 mg/mL). In this case, the cross-sectional diameter of a single RNT was larger and determined to be 8.2 ± 0.5 nm (Figure 3D). Both the TPPO-(G∧C)$_2$ and 6T-(G∧C)$_2$ RNTs were found to be very stable over time with no bundles or precipitates observed in the solutions after one year. Importantly, the observed tubular nanostructures were formed from the self-assembly of G∧C motifs, since control experiments showed that the porphyrin or hexathiophene units alone do not form any well-defined assemblies under the same processing conditions.

The formation of the RNTs led to changes in the UV-Vis spectra (Figure 4A,B). For example, compared to the unassembled TPPO-(G∧C)$_2$ solution, the assembled TPPO-(G∧C)$_2$ RNTs showed significant red shifts in both the porphyrin Soret band and low energy Q band (Figure 4A). More specifically, three new bands appeared at 451 nm, 465 nm and 678 nm, indicating that the peripheral porphyrin groups on the RNTs form J-type aggregates [12]. A solution of 6T-(G∧C)$_2$ RNTs alternatively, showed a 37 nm blue shift (from 419 nm to 382 nm) of the hexathiophene absorption maximum compared to the non-assembled 6T-(G∧C)$_2$ (Figure 4B). This suggests that the peripheral hexathiophene units form H-type aggregates [13]. These absorption differences between the TPPO-(G∧C)$_2$ RNTs and 6T-(G∧C)$_2$ RNTs reflect the different spatial arrangements of the surface functional groups. The porphyrin groups are oriented in a head-to-tail fashion while the hexathiophene units are oriented in a face-to-face fashion.

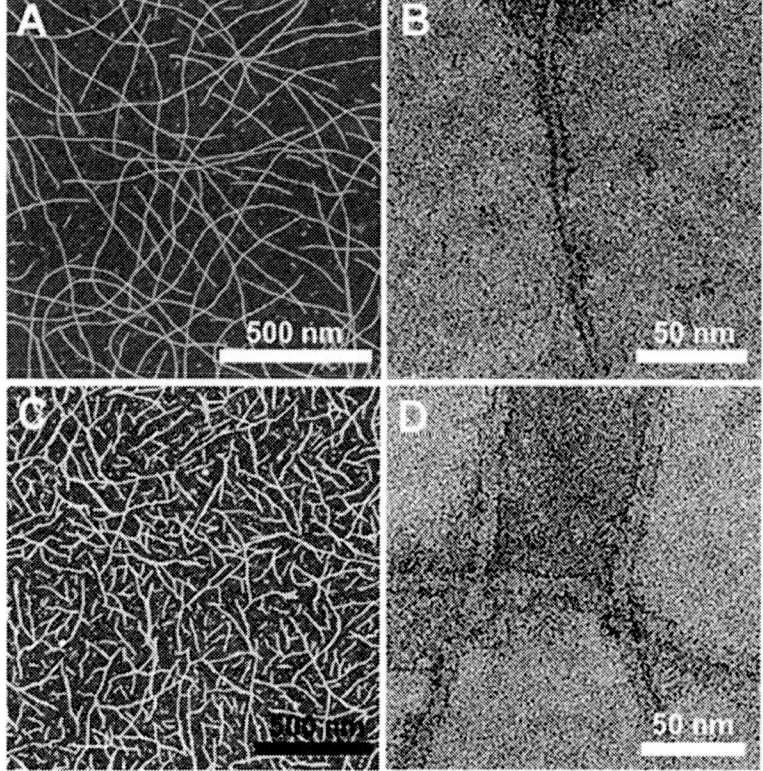

**Figure 3.** SEM and TEM images of (A and B, respectively) TPPO-(G∧C)$_2$ in 1,2-DCB/MeOH, 9:1, v/v and (C and D, respectively) 6T-(G∧C)$_2$ in 1,2-DCB. Concentrations: SEM, 0.1 mg/mL; TEM, 0.01 mg/mL. All samples for imaging were prepared from the corresponding RNT solutions after aging for 24 h. TEM stain: uranyl acetate (0.2% in acetone).

The HOMO energy levels of the TPPO-(G∧C)$_2$ and 6T-(G∧C)$_2$ RNTs were next determined by UPS [14]. The optical band gaps of these materials were calculated from the absorption onsets of the corresponding UV-visible spectra. The LUMO energy levels were then obtained by adding the optical bandgap to the corresponding HOMO energy level (spectra not shown). TPPO-(G∧C)$_2$ RNTs showed a HOMO of -5.62 eV and a LUMO of -4.0 eV and 6T-(G∧C)$_2$ RNTs displayed a HOMO of -5.40 eV and a LUMO of -3.14 eV. The HOMO and LUMO energy levels of PC$_{61}$BM were determined to be -6.10 eV and -4.26 eV, respectively. Thus, the good alignment of the HOMO and LUMO energy levels of the RNTs and PC$_{61}$BM indicates that they are potential D–A pairs for OPVs (Figure 4C).

Interestingly, adding PC$_{61}$BM to the solutions of TPPO-(G∧C)$_2$ or 6T-(G∧C)$_2$ did not affect the nanostructures of the RNTs. In fact, the intact tubular nanostructures were conserved in the mixed D–A solution, as revealed by SEM imaging. This is important for the creation of a well-defined BHJ morphology in solution-processed OPVs. To investigate the efficiency of photoinduced electron transfer from the RNTs to PC$_{61}$BM, solid-state photoluminescence (PL) spectra of spin-cast thin film of TPPO-(G∧C)$_2$ RNTs and the blended film with PC$_{61}$BM (molar ratio 1:1) on quartz were collected. When excited at 446 nm, the film of TPPO-(G∧C)$_2$ RNTs showed strong and broad emission bands at 658 nm and 712 nm in the PL spectrum, which are

4

the characteristic porphyrin emission bands (Figure 4D). In contrast, the emission bands almost disappeared in the blended film of the TPPO-(G∧C)$_2$ RNTs and PC$_{61}$BM, indicating efficient photoinduced electron transfer from the electron donor to the acceptor [15].

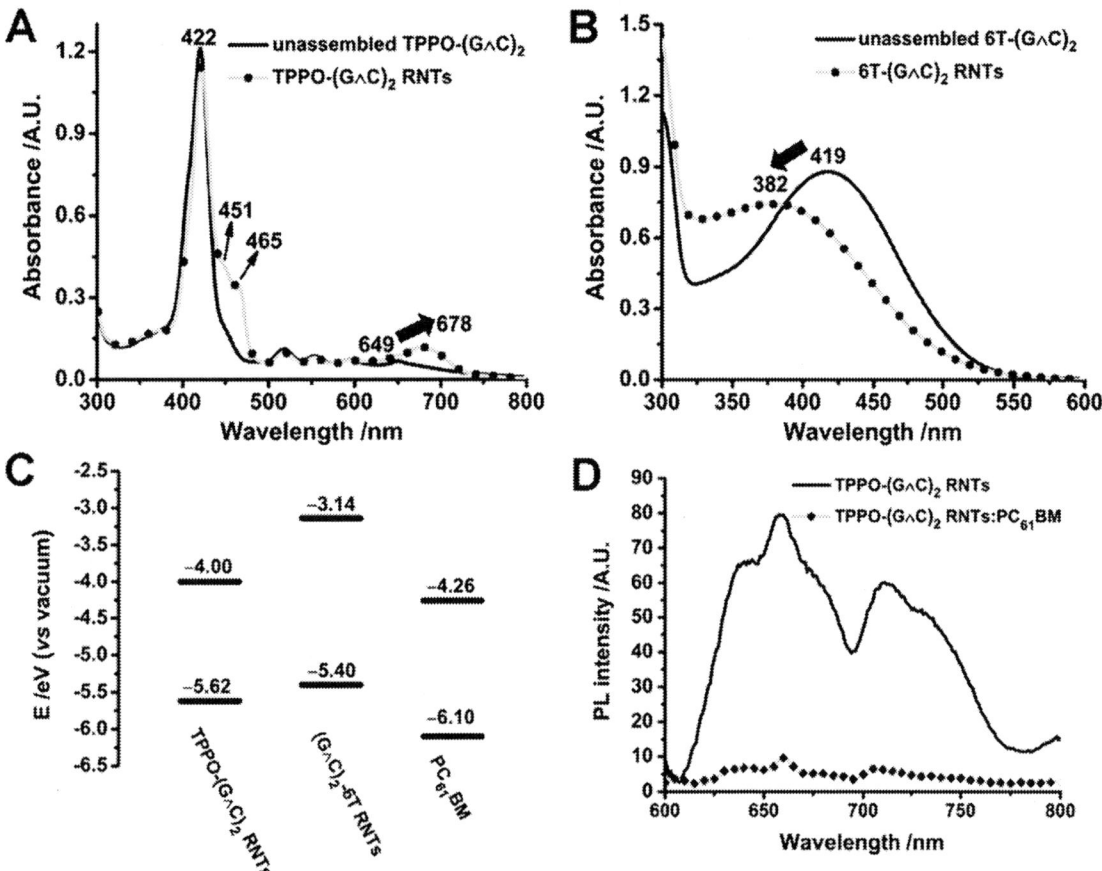

**Figure 4.** UV-Vis spectra of (A) assembled and unassembled TPPO-(G∧C)$_2$ (0.07 mg/mL in 1,2-DCB/MeOH, 9:1, v/v) and (B) assembled and unassembled 6T-(G∧C)$_2$ (0.5 mg/mL in 1,2-DCB); (C) HOMO–LUMO energy levels of TPPO-(G∧C)$_2$ RNTs, 6T-(G∧C)$_2$ RNTs and PC$_{61}$BM; (D) photoluminescence spectra of spin-cast thin film of TPPO-(G∧C)$_2$ RNTs and the blended film with PC$_{61}$BM (molar ratio 1:1) on quartz. Excitation wavelength: 446 nm.

## CONCLUSIONS

Highly ordered porphyrin and hexathiophene-functionalized RNTs were obtained via the hierarchical self-assembly of TPPO-(G∧C)$_2$ and 6T-(G∧C)$_2$ in solution. While the porphyrin groups on the surface of TPPO-(G∧C)$_2$ RNTs were determined to be J-aggregates, the hexathiophene units on the surface of 6T-(G∧C)$_2$ RNTs were shown to be H-aggregates. The good compatibility and well-matched HOMO and LUMO energy levels of the RNTs and PC$_{61}$BM revealed their potential as active-layer materials for solution-processed OPVs.

## ACKNOWLEDGMENTS

We thank National Research Council Canada and the University of Alberta for supporting this project.

## REFERENCES
(*) Corresponding author.

1. B. Kippelen and J.-L. Brédas, *Energy Environ. Sci.* **2**, 251 (2009).
2. A. C. Arias, J. D. MacKenzie, I. McCulloch, J. Rivnay, and A. Salleo, *Chem. Rev.* **110**, 3 (2010).
3. (a) M. T. Dang, L. Hirsch, G. Wantz and J. D. Wuest, *Chem. Rev.* **113**, 3734 (2013). (b) D. Chen, A. Nakahara, D. Wei, D. Nordlund and T. P. Russell, *Nano Lett.* **11**, 561 (2011). (c) P. Vanlaeke, A. Swinnen, I. Haeldermans, G. Vanhoyland, T. Aernouts, D. Cheyns, C. Deibel, J. D'Haen, P. Heremans, J. Poortmans and J. V. Manca, *Sol. Energy Mater. Sol. Cells* **90**, 2150 (2006).
4. (a) F. Würthner, *Nat. Chem.* **6**, 171 (2014). (b) F. J. M. Hoeben, P. Jonkheijm, E. W. Meijer and A. P. H. J. Schenning, *Chem. Rev.* **105**, 1491 (2005).
5. J. A. A. W. Elemans, R. van Hameren, R. J. M. Nolte and A. E. Rowan, *Adv. Mater.* **18**, 1251 (2006).
6. (a) T. Hasobe, K. Saito, P. V. Kamat, V. Troiani, H. Qiu, N. Solladié, K. S. Kim, J. K. Park, D. Kim, F. D'Souza and S. Fukuzumi, *J. Mater. Chem.* **17**, 4160 (2007). (b) R. J. Kumar, J. M. MacDonald, T. B. Singh, L. J. Waddington and A. B. Holmes, *J. Am. Chem. Soc.* **133**, 8564 (2011). (c) L. Schmidt-Mende, A. Fechtenkötter, K. Müllen, E. Moons, R. H. Friend and J. D. MacKenzie, *Science* **293**, 1119 (2001).
7. (a) H. Fenniri, P. Mathivanan, K. L. Vidale, D. M. Sherman, K. Hallenga, K. V. Wood and J. G. Stowell, *J. Am. Chem. Soc.* **12**, 3854 (2001). (b) H. Fenniri, B.-L. Deng and A. E. Ribbe *J. Am. Chem. Soc.* **124**, 11064 (2002).
8. (a) J. Moralez, J. Raez, T. Yamazaki, R. Motkuri, A. Kovalenko and H. Fenniri *J. Am. Chem. Soc.* **127**, 8307 (2005). (b) U. D. Hemraz, M. EI-Bakkari, T. Yamazaki, J.-Y. Cho, R. L. Beingessner and H. Fenniri, *Nanoscale* **6**, 9421 (2014).
9. (a) G. Tikhomirov, M. Oderinde, D. Makeiff, A. Mansouri, W. Lu, F. Heirtzler, D. Y. Kwok and H. Fenniri, *J. Org. Chem.*, **73**, 4248 (2008). (b) R. L. Beingessner, B.-L. Deng, P. E. Fanwick and H. Fenniri, *J. Org. Chem.*, **73**, 931 (2008). (c) R. L. Beingessner, J. A. Diaz, U. D. Hemraz and H. Fenniri, *Tet. Lett.*, **52**, 661 (2011).
10. (a) R. Chhabra, J. G. Moralez, J. Raez, T. Yamazaki, J.-Y. Cho, A. J. Myles, A. Kovalenko and H. Fenniri *J. Am. Chem. Soc.* **132**, 32 (2010).
11. (a) S. S. Suri, F. Rakotondradany, A. J. Myles, H. Fenniri and B. Singh, *Biomaterials* **30**, 3084 (2009). (b) W. S. Journeay, S. S. Suri, J. G. Moralez, H. Fenniri and B. Singh, *Small*, **5**, 1446 (2009).
12. S. Okada and H. Segawa, *J. Am. Chem. Soc.* **125**, 2792 (2003).
13. D. A. Stone, A. S. Tayi, J. E. Goldberger, L. C. Palmer and S. I. Stupp, *Chem. Commun.* **47**, 5702 (2011).
14. B. W. D'Andrade, S. Datta, S. R. Forrest, P. Djurovich, E. Polikarpov and M. E. Thompson, *Org. Electron.* **6**, 11 (2005).
15. M. G. Walter, A. B. Rudine and C. C. Wamser, *J. Porphyrins Phthalocyanines* **14**, 759 (2010).

**Mater. Res. Soc. Symp. Proc. Vol. 1737 © 2015 Materials Research Society**
**DOI: 10.1557/opl.2015.501**

# Computational modeling of nanosecond time-scale charge carrier dynamics in organic semiconductors

Brian Johnson[1], Keshab Paudel[1], Oksana Ostroverkhova[1]

[1]Oregon State University, Corvallis, OR, United States

## ABSTRACT

We present a study of photoinduced charge carrier dynamics in single crystals and polycrystalline thin films of a functionalized fluorinated anthradithiophene (ADT) derivative, ADT-TES-F, combining measurements of time-resolved photocurrent with computational modeling. Simulations revealed two competing charge generation pathways: ultrafast charge separation and nanosecond (ns) time-scale exciton dissociation. Single crystals exhibited significantly enhanced fast charge photogeneration and charge carrier mobilities, as well as lower charge trap densities and free hole-trapped electron recombination, as compared to thin films. At sub-ns time scales after photoexcitation, the light intensity dependence of the photocurrents obtained in single crystals was determined by the carrier density-dependent recombination. At longer time scales, and at lower intensities, taking into account carrier concentration-dependent mobility improved agreement between numerically simulated and experimentally measured photocurrent data.

## INTRODUCTION

Organic semiconductors are of interest for (opto)electronic applications due to their ease of fabrication, low cost, and tunable properties. A considerable research effort has been applied to characterizing charge photogeneration and transport, as well as structure-property relationships, in small-molecule and polymeric materials.[1,2] One promising class of organic materials is solution-processable small-molecule organic semiconductors with high charge carrier mobilities and strong photoresponse. Examples of such materials are functionalized anthradithiophene (ADT) derivatives which display fast photoresponse, thin-film transistor (TFT) charge carrier (hole) mobilities of ~1.5 cm$^2$/Vs in spin-cast films, and high photoconductive gains under continuous wave excitation.[3,4] However, understanding of mechanisms of enhanced charge photogeneration and charge transport in these materials is lacking and is necessary for guiding design of improved materials. To address these issues, we developed in our previous work a numerical model which allowed us to analyze photocurrent dynamics in polycrystalline thin films of functionalized ADT derivatives[5–7] and bulk heterojunctions with the ADT donor. In particular, we quantified the contributions of various charge generation pathways and of charge carrier transport, trapping, and recombination properties to experimentally measured time-resolved photocurrents (TPCs). Here, we extend this work to single crystal devices, which are good model systems for studies of *intrinsic* properties of the material, and compare various parameters pertaining to photoinduced charge carrier dynamics in single crystals and thin films of a fluorinated ADT derivative functionalized with triethylsilylethynyl (TES) side groups, ADT-TES-F (Fig. 1b, inset).

One of the tests of robustness for a model describing photocurrent dynamics is the model's ability to predict dependence of the charge carrier dynamics on various external parameters such as electric field, light intensity, and temperature. Previously, we modeled the electric field dependence of the TPC dynamics which result from that of the charge photogeneration efficiency and charge carrier mobility.[5] In this paper, we focus on the origin of *light intensity-dependent* TPC and, in particular, explore contributions of charge carrier density-dependent recombination (such as bimolecular recombination) and mobility to the TPC amplitude and dynamics. The dependence of mobility on charge carrier concentration (defined as the ratio of free charge carrier and total hopping site densities)[8] has primarily been explored in field-effect transistors, where the carrier concentration can be controlled by varying the gate voltage.[9] However, the field-effect mobility is strongly affected by charge trapping and metal-organic interface effects, which obscure the direct effect of carrier concentration on charge transport. Carrier concentration dependence in the low-mobility ($\mu < 10^{-3}$ cm$^2$/(Vs)) polymers with hopping charge transport has been extensively studied theoretically.[8] A review of several numerical models found that the carrier density dependence of mobility could be parametrized by a simple exponential.[8] These models utilize a Miller-Abrahams's hopping rate and a Gaussian density of states (DOS), the applicability of which to materials with higher mobilities ($\mu > \sim 1$ cm$^2$/Vs) such as organic single crystals, is still under debate[9-13] and is explored here.

## THEORY

A detailed overview of the model used, which describes TPC dynamics in both pristine materials and their donor-acceptor (D/A) blends, can be found elsewhere.[5-7] Briefly, we solve a system of space-averaged coupled differential equations to determine the time evolution of free and trapped charge carrier densities and of exciton density under pulsed light excitation. The model incorporates multiple charge generation paths, electron and hole trapping, hole detrapping, trap assisted recombination, and bimolecular recombination. In pristine materials, the two charge generation paths considered are ultrafast formation of spatially separated carriers (SSC) with efficiency $\xi_{SSC}$ and formation (with efficiency $\xi_{FE}$) and dissociation of Frenkel excitons (FEs) during their lifetimes $\tau$ (typically >10 ns) with efficiency $\eta$; these were previously quantified in pristine films of ADT-TES-F (inset of Fig.1(b)) and its D/A blends.[5-7] Here we explore (i) the differences between time-resolved charge carrier dynamics in film and single crystal and (ii) the light intensity-dependent photocurrent dynamics in single crystals. For (ii), we tested two hypotheses: charge carrier density-dependent recombination and charge carrier density-dependent mobility, respectively, as the dominant factor. For that, we have modified the temperature ($T$)- and electric field ($E$)-dependent Poole-Frenkel charge carrier mobility[5-7] $\mu_{p(n)0}(T,E)$ to incorporate a carrier density dependence as follows:

$$\mu_{p(n)} = \mu_{p(n)0}(T,E) * f_m\left(p(n)_f, T\right) \tag{1}$$

$$f_1\left(p(n)_f\right) = 1 \tag{2}$$

$$f_2\left(p(n)_f\right) = exp\left(u_{p(n)}\left(2p(n)_f / N_t\right)^{v_{p(n)}}\right) \tag{3}$$

where $p(n)_f$ is the free hole (electron) density and $f_m\left(p(n)_f, T\right)$ is a carrier density dependence function (Eqs.(2) and (3)). The function $f_1$ (Eq. (2)) assumes no carrier density dependence of mobility, previously predicted for the case of carrier concentrations of $10^{-4}$-$10^{-6}$ carriers/site, similar to ours.[14] Therefore, the light intensity-dependent TPC dynamics simulated using $f_1$ would be due to carrier density-dependent recombination only. The function $f_2$ (Eq.(3)) is a parametrization of numerical hopping conduction results derived by Coehoorn et al.,[8] which assumes a Gaussian DOS with width $\sigma_{p(n)}$ for hole(electron) conduction. $N_t$ is the number of possible hopping sites (taken to be $10^{20}$ cm$^{-3}$), $u_{p(n)} = (1/2)(\hat{s}_{p(n)}^2 - \hat{s}_{p(n)})$, $v_{p(n)} = (2/\hat{s}_{p(n)}^2)( \ln(\hat{s}_{p(n)}^2 - \hat{s}_{p(n)}) - \ln(\ln(4)))$, $\hat{s}_{p(n)} = \sigma_{p(n)}/k_B T$. In this approach, charge carrier mobility is an increasing function of carrier concentration, which in our experiments is light intensity-dependent. Therefore, in this case the TPC incorporates intensity-dependent contributions of mobility and of various mobility-dependent rates.[5] In our model, the mobility-dependent rates are bimolecular recombination and FE dissociation; both rates have a linear dependence on mobility.[5]

Using carrier density-dependent mobility with function $f_2$ also requires a change in how initial conditions are calculated. In particular, we must now numerically solve the following equations to determine initial conditions (i.e. in the dark, before photoexcitation) for holes and electrons as a system:[5]

$$n_f^0 e^{\frac{\varphi_B^n}{k_B T}} \left( \mu_{n0} f_m\left(n_f^0, T\right) e^{\frac{-\varphi_B^n}{k_B T}} + \mu_{p0} f_m\left(p_f^0, T\right) e^{\frac{-\varphi_B^p}{k_B T}} \right) = \frac{J_{Dark}}{eE} \tag{4}$$

$$p_f^0 e^{\frac{\varphi_B^p}{k_B T}} \left( \mu_{n0} f_m\left(n_f^0, T\right) e^{\frac{-\varphi_B^n}{k_B T}} + \mu_{p0} f_m\left(p_f^0, T\right) e^{\frac{-\varphi_B^p}{k_B T}} \right) = \frac{J_{Dark}}{eE} \tag{5}$$

where $p(n)_f^0$ is the initial hole (electron) density, $J_{Dark}$ is the measured dark current density at the applied electric field $E$, $k_B$ is Boltzmann's constant, and $\phi_B^{p(n)}$ is the injection barrier for holes (electrons), taken to be 0.25 (2.05) eV.

## RESULTS AND DISCUSSION

Polycrystalline films of ADT-TES-F were spin cast from chlorobenzene solution onto pentafluorobenzenethiol (PFBT)-treated interdigitated Au electrodes with a 25 μm gap deposited on glass substrates. Similar substrates were used for single crystal growth, for which a 30 mM chlorobenzene solution of ADT-TES-F was deposited on the substrate which was then placed in the refrigerator. Such procedure yielded >100 μm size crystals, which were then separated and characterized using polarization microscopy. The samples were excited with a 470 ps 532 nm laser, and the TPCs were measured using a 50 GHz digital sampling oscilloscope, as described in detail in our previous publications.[5,7] The time resolution was about 0.5 ns, limited by the laser pulse width. The incident pulse fluence was varied using a Thorlabs neutral density filter wheel in the 2 – 39 μJ/cm$^2$ range. The experimental data were then fit with numerically simulated photocurrents, and various parameters of the model were determined as detailed elsewhere.[5]

### Single crystal versus thin film

The examples of simulated photocurrents obtained for the thin film and single crystal devices using $f_1$ and the parameter values reported in Table I (col 1 & 3) are presented in Fig. 1. Table I also includes results from a fit of single crystal TPC at 20 kV/cm and 5 μJ/cm$^2$. The single crystal devices exhibited considerably higher peak photocurrent densities and slower initial decay dynamics as compared to those in thin film samples (Fig.1). The ultrafast SSC charge generation pathway (characterized by $\xi_{SSC}$, Table I) was significantly more efficient in single crystals than in films, consistent with previous work showing a strong dependence of SSC efficiency on film crystallinity.[7] In contrast, the FE dissociation-based contribution to charge generation ($\xi_{FE}*\eta$) was considerably reduced in the single crystal, which could be related to a considerably higher FE recombination rate $k_R$ as compared to that in films (Table I). The rate of the initial photocurrent decay depends on charge carrier recombination and trapping properties; for example, faster initial decay in ADT-TES-F/PCBM blends as compared to pristine ADT-TES-F films has previously[7] been attributed to bimolecular recombination. A considerably slower initial decay rate in single crystal devices as compared to thin films (Fig. 1) is a combination of reduced trap densities ($N_p$ and $N_n$), shallower traps ($\Delta$), and reduced free hole-trapped electron recombination ($B_{p_f n_t}$) in the single crystal (Table I).

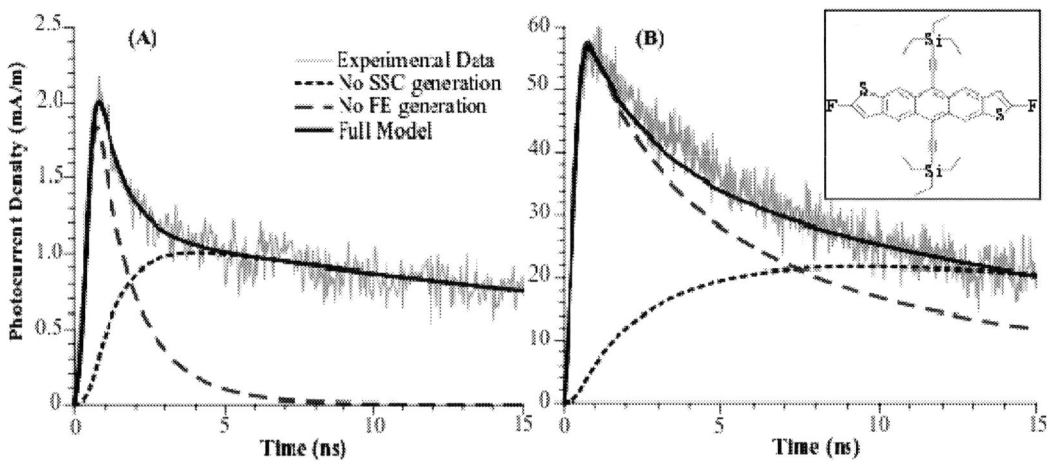

**Figure 1.** Experimental (grey) and simulated (black) TPC densities for (A) a thin film at 50 kV/cm and 2 µJ/cm² and (B) a single crystal at 20 kV/cm and 24 µJ/cm². Other lines show simulations run with $\xi_{SSC} = 0$ (dotted line) and $\xi_{FE} = 0$ (dashed line). Inset shows the molecular structure of ADT-TES-F.

**Table I.** Selected parameter values extracted from fits of experimental data to the numerical model[5] using $f_1$. Column 3 used a similar procedure to our previous voltage dependent fits.[5]

| Sample: | 1. Thin Film 2 µJ/cm² | 2. Single Crystal 5 µJ/cm² | 3. Single Crystal 5-39 µJ/cm² |
|---|---|---|---|
| $\mu_{n0}(T,E)$ (cm²/Vs) | 0.2 | 0.7 | 0.7 |
| $\mu_{p0}(T,E)$ (cm²/Vs) | 0.9 | 2.0 | 2.1 |
| $\xi_{SSC}$ | 0.04 | 0.22 | 0.24 |
| $\xi_{FE}*\eta$ | 0.33 | 0.05 | 0.05 |
| $k_{Diss,FE}$ (s⁻¹) | $2.6 \times 10^7$ | $1.7 \times 10^7$ | $1.7 \times 10^7$ |
| $k_R$ (s⁻¹) | $1.0 \times 10^7$ | $7.6 \times 10^7$ | $7.6 \times 10^7$ |
| $N_n$ (cm⁻³) | $3.0 \times 10^{19}$ | $3.2 \times 10^{18}$ | $3.5 \times 10^{18}$ |
| $N_p$ (cm⁻³) | $4.4 \times 10^{18}$ | $2.7 \times 10^{18}$ | $3.7 \times 10^{18}$ |
| $B_{n_f p_t}$ (cm³/s) | $1.9 \times 10^{-3}$ | $4.8 \times 10^{-3}$ | $4.0 \times 10^{-3}$ |
| $B_{p_f n_t}$ (cm³/s) | $2.1 \times 10^{-6}$ | $1.6 \times 10^{-7}$ | $9.4 \times 10^{-8}$ |
| $\Delta$ (meV) | 31 | 20 | 15 |

**Light intensity dependence of photocurrent dynamics**

Comparisons between TPC fits using $f_1$ and $f_2$ functions (Eqs. (2) & (3)) at multiple incident intensities in the single crystal sample are shown in Figure 2. Although both $f_1$ and $f_2$ approaches could reproduce the data reasonably well at particular light intensities (e.g. Fig. 2A), neither $f_1$ nor $f_2$ fully captured the TPC dynamics over the entire light intensity range studied (Fig. 2B-D), which corresponded to carrier concentrations of $10^{-5}$-$10^{-4}$ carriers/site. Figures 2C and 2D show TPCs integrated over a 20 ns time period and TPC peak amplitudes, respectively, obtained from experimental data (circles) and numerically simulated TPCs using $f_1$ (squares) and $f_2$ (diamonds). The $f_1$ approach slightly overestimated the TPC amplitude throughout the entire intensity range and overestimated the extracted charge at low intensities. The $f_2$ approach predicted a stronger intensity dependence of the TPC amplitude than the one observed experimentally, but matched the change in the extracted charge with intensity better than $f_1$ throughout the entire intensity range. This indicates that light intensity dependence of the TPC dynamics at sub-ns time scales (reflected in that of the TPC amplitude) is mostly determined by charge carrier density-dependent recombination (bimolecular and/or trap-assisted), in which mobility does not depend on carrier concentration. However, at longer time scales (reflected in the integrated TPC), and at low intensities (<10 µJ/cm$^2$), additional intensity-dependent contributions may become

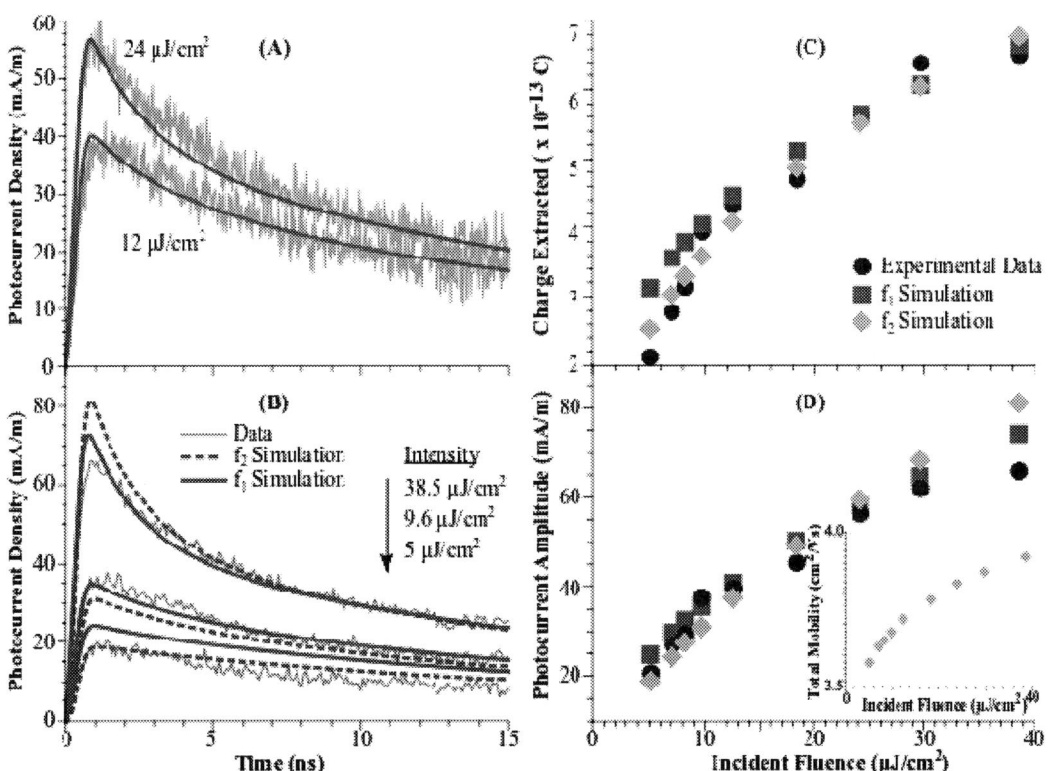

**Figure 2.** (A) TPC data along with simulations using $f_1$ for two light intensities. (B) TPC data and simulations using $f_1$ and $f_2$ for multiple intensities. (C) Integrated TPC photocurrent values and (D) TPC amplitude for experiment and simulations using $f_1$ and $f_2$. Inset: Sum of mobilities ($\mu_n + \mu_p$) calculated using $f_2$ for multiple intensities.

important. This includes carrier density-dependent charge carrier mobility, for which the simulations using the $f_2$ approach predicted about a ~10% increase in an overall mobility in the range of light intensities used in our experiments (Fig. 2D, inset).[8]

## CONCLUSIONS

We applied a numerical model to an analysis of experimentally measured time-resolved photocurrent dynamics[5,7] in ADT-TES-F single crystal and thin film devices. Contributions of several charge photogeneration pathways to the photocurrent were quantified, and parameters describing charge transport, trapping, and recombination properties were determined. In single crystals, we observed large enhancements in the ultrafast charge carrier separation efficiency and in charge carrier mobilities, as well as reduction in charge trap densities and in free hole-trapped electron recombination, as compared to thin films. In single crystals at sub-ns time scales after photoexcitation, the light intensity dependence of the photocurrent is largely determined by the carrier density-dependent recombination. At longer time scales, and at lower intensities, taking into account carrier concentration-dependent mobility improves agreement between numerically simulated and experimentally measured photocurrent data.

## ACKNOWLEDGMENTS

We thank Prof. J.E. Anthony for ADT-TES-F and Prof. O. D. Jurchescu for single crystal samples for our initial studies and for helpful tips on the single crystal growth. This work was supported by the NSF grant DMR-1207309.

## REFERENCES

[1] T. Clarke and J. Durrant, Chem. Rev. **110**, 6736 (2010).

[2] B. Walker, C. Kim, and T.-Q. Nguyen, Chem. Mater. **23**, 470 (2011).

[3] R.J. Kline, S.D. Hudson, X. Zhang, D.J. Gundlach, A.J. Moad, O.D. Jurchescu, T.N. Jackson, S. Subramanian, J.E. Anthony, M.F. Toney, and L.J. Richter, Chem. Mater. **23**, 1194 (2011).

[4] A.D. Platt, J. Day, S. Subramanian, J.E. Anthony, and O. Ostroverkhova, J. Phys. Chem. C **113**, 14006 (2009).

[5] B. Johnson, M.J. Kendrick, and O. Ostroverkhova, J. Appl. Phys. **114**, 094508 (2013).

[6] B. Johnson, K. Paudel, M.J. Kendrick, and O. Ostroverkhova, Proc. SPIE **8830**, 88301S (2013).

[7] K. Paudel, B. Johnson, M. Thieme, M.M. Haley, M.M. Payne, J.E. Anthony, and O. Ostroverkhova, Appl. Phys. Lett. **105**, 043301 (2014).

[8] R. Coehoorn, W. Pasveer, P. Bobbert, and M. Michels, Phys. Rev. B **72**, 155206 (2005).

[9] H. Sirringhaus, T. Sakanoue, and J.-F. Chang, Phys. Stat. Sol. B **249**, 1655 (2012).

[10] H. a v Laarhoven, C.F.J. Flipse, M. Koeberg, M. Bonn, E. Hendry, G. Orlandi, O.D. Jurchescu, T.T.M. Palstra, and A. Troisi, J. Chem. Phys. **129**, 044704 (2008).

[11] Y.C. Cheng, R.J. Silbey, D. A. da Silva Filho, J.P. Calbert, J. Cornil, and J.L. Brédas, J. Chem. Phys. **118**, 3764 (2003).

[12] J. Cottaar, L. Koster, R. Coehoorn, and P. Bobbert, Phys. Rev. Lett. **107**, 136601 (2011).

[13] A. Troisi, J. Chem. Phys. **134**, 034702 (2011).

[14] M.E. Gershenson, V. Podzorov, and A.F. Morpurgo, Rev. Mod. Phys. **78**, 973 (2006).

**Mater. Res. Soc. Symp. Proc. Vol. 1737 © 2015 Materials Research Society**
**DOI: 10.1557/opl.2015.502**

# The Influence of Morphology on the Charge Transport in Two-Phase Disordered Organic Systems

Cristiano F. Woellner[1], Leonardo D. Machado[1], Pedro A. S. Autreto[1], José A. Freire[2] and Douglas S. Galvão[1]

[1]Applied Physics Department, State University of Campinas, 13083-970 Campinas, São Paulo, Brazil.

[2]Physics Department, Federal University of Paraná, 81531-990, Curitiba, Paraná, Brazil

## ABSTRACT

In this work we use a three-dimensional Pauli master equation to investigate the charge carrier mobility of a two-phase system, which can mimic donor-acceptor and amorphous-crystalline bulk heterojunctions. Our approach can be separated into two parts: the morphology generation and the charge transport modeling in the generated blend. The morphology part is based on a Monte Carlo simulation of binary mixtures (donor/acceptor). The second part is carried out by numerically solving the steady-state Pauli master equation. By taking the energetic disorder of each phase, their energy offset and domain morphology into consideration, we show that the carrier mobility can have a significant different behavior when compared to a one-phase system. When the energy offset is non-zero, we show that the mobility electric field dependence switches from negative to positive at a threshold field proportional to the energy offset. Additionally, the influence of morphology, through the domain size and the interfacial roughness parameters, on the transport was also investigated.

## INTRODUCTION

High performance conjugated polymers have gained significant attention in recent years due to their low-cost processing and high ductility, fundamental features for applications in flexible electronics. These features provide them with a significant competitive advantages over other technologies based on crystalline semiconductors [1]. However, despite their great potential for applications, weak device performance is still a limiting factor and many fundamental questions regarding the origin of these problems remain unsolved. Thus, a better understanding of charge transport in these materials is necessary in order to overcome these difficulties.

With the advent of solar cells based on bulk heterojunctions (BHJs) [2,3] the need for models that describe the charge transport in such systems increased. In 2005 Watkins *et al.* [4] proposed an effective model based on the dynamical Monte Carlo method to generate the morphologies of binary mixtures, and in 2010, using a proper combination of Pasveer [5] and Watkins approaches, Koster [6] showed for the first time that the mobility in donor-acceptor blends could exhibit a negative electric field dependence.

Despite these advances, all models so far for charge transport in BHJs explicitly assumed that the transport occurs exclusively in one phase (donor or acceptor), independently of the difference in energy between the electronic states of the two phases or on the applied electric field value [7].

**THEORY**

In this work we investigated the charge carrier mobility of a two-phase system using an approach that can be separated into two parts: the morphology generation and the charge transport modeling in the generated blend. The morphology part is based on a lattice-gas model of a binary mixture developed by Watkins *et al.* [4]. The system is defined on a regular cubic lattice of $N$ sites and lattice parameter $a$. The phase-1 is constituted by $\alpha N$ sites and phase-2 by $(1-\alpha)N$ sites, where $\alpha$ is the volume ratio. The initial lattice configuration is a random mixture of the constituents with a fixed $\alpha$. In order to simulate a realistic two-phase system, such as donor-acceptor or amorphous-crystalline blends, phase segregation is induced. This is accomplished thorough Monte Carlo simulation by adjusting the interaction energy between the constituents. At every Monte Carlo step a pair of neighboring sites is randomly chosen and the total energy of the system before and after the sites swap their positions is calculated. If the total energy decreases the swap is automatically accepted, otherwise a non-zero probability of acceptance is associated with the exchange.

The phase separation is characterized by two parameters: a characteristic length, known as the domain size, defined by [6,8]:

$$b = \frac{6(1-\alpha)V}{A}, \tag{1}$$

where $V$ is the total volume and $A$ is the interfacial area. The second parameter is associated with the interfacial roughness between the two phases, named the interfacial roughness parameter, $\gamma$. The $\gamma$ values are defined controlling the interaction energy between the sites (black/white, see Figure 1) of which phase. Lower $\gamma$ values imply high roughness values. Figure 1 illustrates two typical cases: a rougher interfacial surface ($\gamma=0.1$) and a smoother one ($\gamma=0.7$). The Monte Carlo simulation is continuously carried out until the desired domain size is reached.

**Figure 1**. Two representative morphologies with an equal amount of constituents (volume ratio $\alpha=0.5$) and the same domain size (b=6 nm). The left case illustrates a rougher interfacial surface ($\gamma=0.1$) and the right one illustrates a smoother surface ($\gamma=0.7$).

Disordered organic materials, due to the large morphological disorder and the weak electronic coupling, have localized electronic states. The energetic distribution of these localized states in a one-phase system is usually assumed to be Gaussian [9]. In the present work we will consider just the electrons. For the two-phase case it can be assumed a bimodal Gaussian density of states, where each phase is characterized by a Gaussian distribution with width $\sigma_1$ (phase-1) and $\sigma_2$ (phase-2). Hereafter, we will assume $\sigma_1 = \sigma_2 = 0.1$ eV.

The charge carrier mobility is calculated by numerically solving the steady-state Pauli master equation:

$$\sum_{j \neq i} [W_{i \to j} P_i (1 - P_j) - W_{j \to i} P_j (1 - P_i)] = 0, \tag{2}$$

where $P_i$ is the probability that site $i$ is occupied by a charge carrier, and $W_{i \to j}$ is the hopping rate from site $i$ to site $j$. The $(1-P_i)$ term excludes, in a mean-field approximation, the possibility of double occupancy. The hopping rate is assumed to be of the Miller-Abrahams form and the electric field, $F$, is assumed constant through the system.

We solved Eq. (2) for the occupational probabilities, $P_i$'s, using periodic boundary conditions by an iteration procedure and once the occupational probabilities are obtained, the charge-carrier mobility $\mu$ is calculated from:

$$\mu = \frac{\sum_{i, j \neq i} W_{i \to j} P_i (1 - P_j) R^x}{nFV} \tag{3}$$

where $n = <P_i>/a^3$ and $V$ is the system volume. The lattice size is $N=100^3$. Averages over a number of different disorder configurations and different morphologies (with fixed domain size, interfacial roughness and volume ratio) were taken until accuracy better than 10% was obtained for $\mu$. Throughout the paper we used fixed values for the carrier density ($n=10^{-5}a^{-3}$), the lattice constant ($a=1$ nm), and the thermal energy ($k_B T=0.025$ eV, corresponding to room temperature).

**DISCUSSION**

In figure 2 we show, using a linear scale, the effect of the domain size, $b$, and the interfacial roughness, $\gamma$, on the mobility, normalized to the "$b=6$ nm" and "$\gamma=0.7$" mobility $\mu_0=1.12 \times 10^{-4}$ cm$^2$/Vs. For $\gamma$ larger than some threshold interfacial roughness parameter (here $\gamma >= 0.2$) we can clearly seen from the Figure that increasing $b$ also enhances the mobility. The domain size $b$ has a direct impact on the channel network of each phase. As it increases the number of direct percolative paths from one electrode to the other, consequently this increases the mobility. But for a rougher interfacial surface, $\gamma < 0.2$, the carrier mobility does not increases monotonically but has a maximum value at a specific domain size value. This means that the optimal value for the carrier mobility is achieved by a right combination of both domain size and interfacial roughness parameters.

## ELECTRIC FIELD DEPENDENCE

We have discussed in the last section the effect of morphology in the zero field regime. In this section we will discuss the influence of the electric field on the mobility for the case presented in the previous section: both phases equally disordered. We will assume here $b$=6 nm and $\alpha$=0.5.

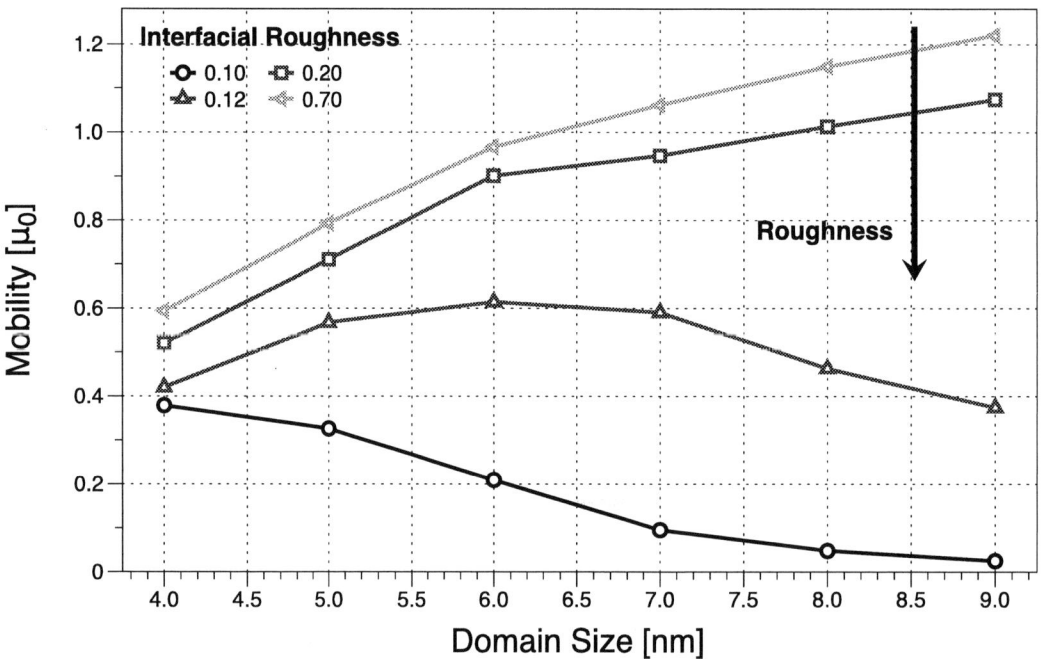

**Figure 2**. Mobility values as a function of the domain size, $b$, for selected values of the interfacial roughness, $\gamma$. For all the investigated cases we used the volume ratio $\alpha$=0.5, and both phases are equally disordered, $\sigma_1$= $\sigma_2$=0.1 eV. The mobility is normalized to the "$b$=6 nm" mobility $\mu_0$=1.12x10$^{-4}$ cm$^2$/Vs.

In Figure 3 we show, in a semi-log scale, the effect of the electric field, $F$, on the mobility values. We fixed $\sigma_1$=$\sigma_2$=0.1 eV and considered four different values of $\gamma$. For $E_{offset}$ >0 (here $E_{offset}$=0.1 eV), the mobility has a negative field dependence followed by positive one [10]. The negative field dependence was discussed by Bässler [9] and Koster [6] in the context of a single-phase system. The effect was explained in terms of paths that go against the field that contribute to the mobility at very low fields but cease to contribute (hence decreasing the mobility) as the field increases. Above a certain $F_{min}$ only paths that go mostly along the field contribute and the mobility along these paths increases with increasing $F$. What is displayed in figure 3 is a somewhat similar effect on a two-phase system. For $F$ lower than some threshold $F_{min}$ the carrier is entirely restricted to the less energetic phase-2 at low fields, the observed decrease of $\mu$ below $F_{min}$ is the effect of the field in the transport through the channel network of phase-2. Above $F_{min}$ the field can provide energy for the carrier to hop into the more energetic phase-1 (it is significant that e$F_{min}a$ is of the order of $E_{offset}$), and a number of paths that were forbidden at low fields become available, resulting in a mobility increase. Finally, Figure 3 also shows that the roughness affect is more pronounced for higher electric field values.

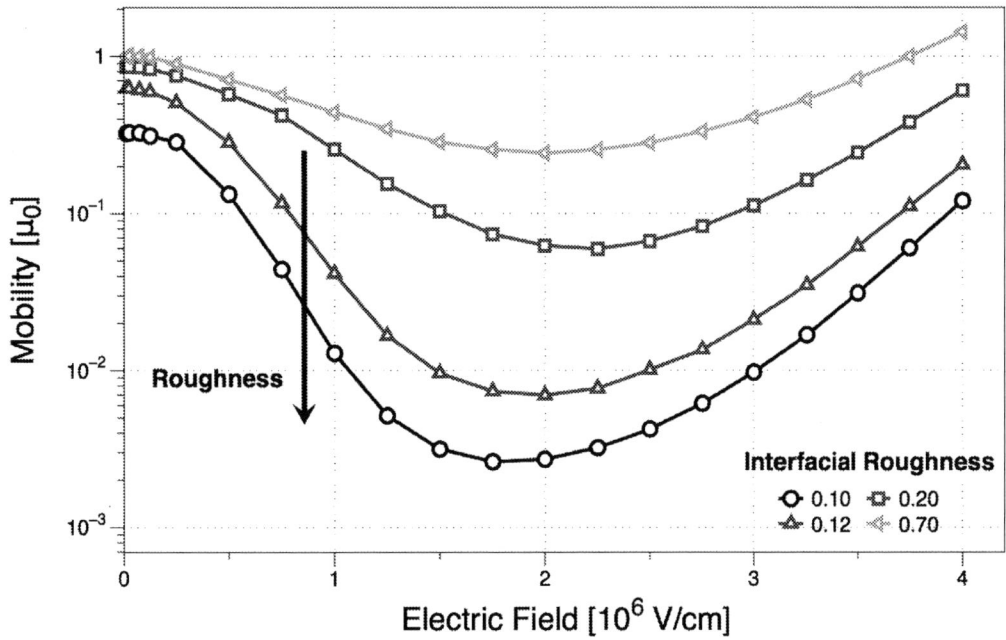

**Figure 3**. Mobility values as a function of the electric field, for selected values of the interfacial roughness $\gamma$. For all cases, the domain size was $b$=6 nm, the volume ratio $\alpha$= 0.5, and both phases are equally disordered, $\sigma_1=\sigma_2=0.1$ eV. The mobility values are normalized to the zero-field "$b$=6 nm" mobility $\mu_0=1.12 \times 10^{-4}$ cm$^2$/Vs. We can clearly see that the roughness strongly affects the mobility values.

**CONCLUSIONS**

We have applied a three-dimensional Pauli master equation model with a bimodal Gaussian density of states to investigate the influence of morphology and electric field on the charge carrier mobility of a two-phase system in the low carrier density limit. At low electric fields and the two phases being equally disordered, we showed that the carrier mobility have a completely different behavior depending on the domain size, $b$, and the interfacial roughness, $\gamma$, parameters. For the same domain size, the carrier mobility decreases with increasing the roughness. For each value of the interfacial roughness parameter the carrier mobility has a maximum value at a specific domain size value. This suggests that the optimal value for the carrier mobility can be obtained by a right combination of both parameters. We have also shown that the electric field dependence on the mobility shows a minimum value which is strongly dependent on the interfacial roughness. In practice, these findings provide in an approximate way, for which regime (negative or positive mobility field dependence) a particular system, for instance a donor-acceptor blend will operate at a given electric field.

**ACKNOWLEDGMENTS**

This work was supported in part by the Brazilian Agencies CAPES, CNPq and FAPESP. The authors thank the Center for Computational Engineering and Sciences at Unicamp for financial support through the FAPESP/CEPID Grant # 2013/08293-7.

**REFERENCES**

1. T. W. Kelley, P. F. Baude, C. Gerlach, D. E. Ender, D. Muyres, M. A. Haase, D. E. Vogel, and S. D. Theiss, *Chem. Mater.* **16**, 4413 (2004).
2. G. Yu, J. Gao, J. C. Hümmelen, F.Wudl, and A. J. Heeger, *Science* **270**, 1789 (1995).
3. P. Blom, V. Mihailetchi, L. J. A. Koster, and D. E. Markov, *Adv. Mater.* **19**, 1551 (2007).
4. P. K.Watkins, A. B.Walker, and G. L. B. Verschoor, *Nano Lett.* **5**, 1814 (2005).
5. W. F. Pasveer, J. Cottaar, C. Tanase, R. Coehoorn, P. A. Bobbert, P.W. M. Blom, D. M. de Leeuw, andM. A. J.Michels, *Phys. Rev. Lett.* **94**, 206601 (2005).
6. L. J. A. Koster, *Phys.Rev.B* **81**, 205318 (2010).
7. C. Deibel and V. Dyakonov, *Rep. Prog. Phys.* **73**, 096401 (2010).
8. Y. Y. Yimer, P. A. Bobbert, and R. Coehoorn, *J. Phys.: Condens. Matter* **20**, 335204 (2008).
9. H. Bässler, *Phys. St. Solidi B* **107**, 9 (1981).
10. C. F. Woellner, Z. Li, J. A. Freire, G. Lu, and T.-Q. Nguyen, *Phys.Rev. B* **88**, 125311 (2013).

**Mater. Res. Soc. Symp. Proc. Vol. 1737 © 2015 Materials Research Society**
**DOI: 10.1557/opl.2015.529**

## Branched Thiophene Oligomer/Polymer Bulk Heterojunction Organic Solar Cell

Francisco Martinez[1], Gloria Neculqueo[1], Sergio O. Vasquez[1], Helge Lemmetyinen[2], Alexander Efimov[2] and Paola Vivo[2]

[1]Dpto Ciencia de Materiales , Facultad de Ciencias Físicas y Matemáticas.Universidad de Chile, Santiago, Chile

[2]Dpt. Bioengineering and Chemical Dept. Tampere University of Technology, Tampere, Finland

### ABSTRACT

Thiophene small novel branched structures have been proposed as candidates for dopant agents transporting holes-electron in organic solar cell (OSC). Low-band gap of these branched oligotiophene have been obtained to be used in organic solar cells. Two branched thiophene oligomers, a sexithienylene vinylene (E)-Bis-1,2-(5,5′′-Dimethyl-(2,2′:3′,2′′-terthiophene) vinylene ,(BSTV) and octathienylene vinylene (BOTV) (E)-Bis-1,2-(5,5′′′-Dimethyl-(2,2′:5′,2′′:3′,2′′′-tetrathiophene) vinylene oligomers, have been synthesized and used as electron donor or dopant in a bulk heterojunction poly(3-hexylthiophene) (P3HT), /[6,6]-phenyl C61-butyric acid methylester (PCBM), Organic Photovoltaic cell .

### INTRODUCTION

Organic photovoltaic (OPV) cells are receiving considerable attention as an attractive alternative to inorganic based photovoltaic devices. Different cells configurations are being used to improve photon absorption, charge separation and transport in organic solar cells [1-6]. Usually, organic solar cells are based either on two superposed layers or on a homogeneous mixture of two organic materials [7]. The homogeneous mixture of two organic materials are based on bulk heterojunction (BHJ) architecture [8], where an electron acceptor (A) molecule is dispersed in an electron donor (D), usually a conjugated polymer [9]. The BHJ allows improving power efficiency by increasing the area of interface between the D and the A, through spontaneous phase segregation of both phases. This continuous interpenetrating network of D-A in the BHJ structure offers a large area for the exciton dissociation and by this way reducing the length of separated excitons to the corresponding electrodes. The BHJ has become a typical structure for OPV improving the performance of the devices. High energy conversion efficiency of the BHJ based solar cells has been drastically improved, where an electron acceptor such as (6,6)-phenyl C61 butyric acid methyl ether (PCBM) and an electron donor such as poly(3-hexylthiophene) (P3HT) are blended to form one mixed film.

Recent improvement of the performance of small molecules OPV cells induces a growing interest for these devices [10]. Small molecules OPV cells based on thiophene oligomers are among the best studied because their very good transport properties, high polarizability as well as their tunable optical and electrochemical properties. The purpose of this work is to contribute with two novel branched thiophene oligomers, BSTV and BOTV which have six and eight thiophene units respectively, highly conjugated, low band gap, soluble and processable structures. In comparison for example, linear sexi or octithiophene, are quite insoluble. In order to work with BHJ architecture, donor and acceptor materials need to be solubilized in a common organic solvent, for this reason our branched oligomers are suitable structures as donor dopants in OSC.

**EXPERIMENTAL**

Synthesis and characterization of (E)-Bis-1,2-(5,5′′-Dimethyl-(2,2′:3′,2′′-terthiophene)vinylene. (BSTV). This compound was synthesized through the McMurry reaction, starting from 5,5′′-dimethyl-2,2′.3′,2′′-terthiophene-5′-carbaldehyde.

The figure 1 shows the scheme synthesis of BSTV

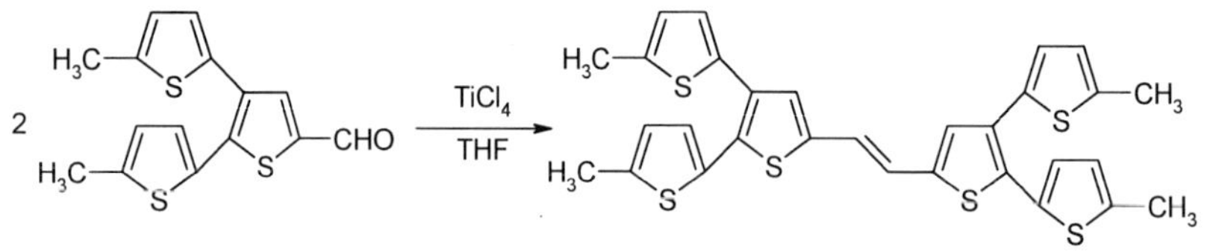

**BSTV**

**Figure 1** Synthesis of (E)-Bis-1,2-(5,5′′-Dimethyl-(2,2′:3′,2′′-terthiophene)vinylene

To a solution of 1g of [10] in 13 mL of anhydrous THF is added 1 mL of titanium tetrachloride at -18°C under Argon atmosphere. The mixture is stirred for half an hour at -18°C. To this solution was added 1.19 g of Zinc powder in small portions for 10-15 min. The mixture was stirred for 40 min. at -18°C and was then gradually warmed to room temperature and then heated at reflux overnight. Then, the mixture was cooled in an ice-water bath and 50 mL of aqueous solution of sodium carbonate 10 % were added. The aqueous layer was extracted with chloroform (3x50 mL). The combined organic layers were filtered and washed with water until neutral pH. The organic layer was dried over anhydrous MgSO$_4$. The product was purified by column chromatography on silica gel using hexane as the eluent, yielding a yellow orange solid. (0.81 g; 86% yield. $^1$H-NMR (CDCl$_3$, 300 MHz) δ (ppm) 7.02 (2H, s), 6.97 (2H, d, J=3.5 Hz), 6.92 (2H, s), 6.89 (2H, d, J=3.5 Hz), 6.67 (4H, m), 2.49 (12H, m). $^{13}$C-NMR (CDCl$_3$, 75 MHz) δ (ppm) 141.5, 140.2, 139.8, 134.8, 132.8, 132.0, 130.9, 129.3, 127.6, 126.7, 125.5, 125.3, 121.3, 15.47, 15.43, HRMS (ESI-TOF; CHCl$_3$/MeOH 10:15); m/z 576.0210. [M+H]$^{+(}$ calcd for C$_{30}$H$_{24}$S$_6^+$ 576.9071).

Synthesis and characterization of (E)-Bis-1,2-(5,5′′-Dimethyl-(2,2′:3′,2′′:3′,2′′′′-tetrathiophene)vinylene (BOTV). The figure 2 shows the scheme synthesis of BOTV

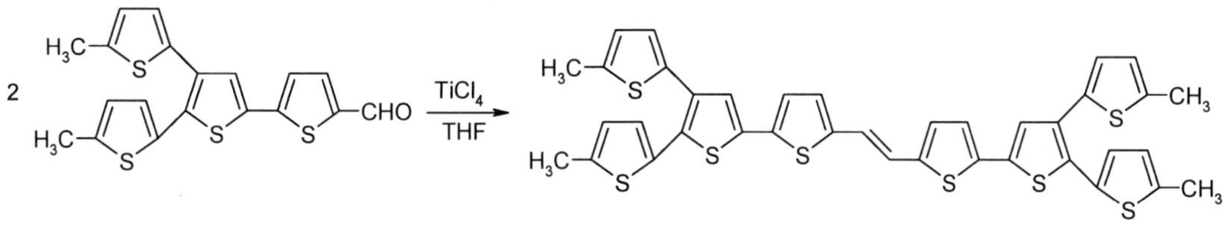

**Figure 2** Synthesis of (E)-Bis-1,2-(5,5″-Dimethyl-(2,2′:3′,2″:3',2'''-tetrathiophene) vinylene (BOTV)

This compound was synthesized through the McMurry reaction in a similar manner as described above for BSTV, this time the starting compound was 5,5'''-dimethyl-2,2′:5′,2″:3′,2'''-tetrathiophene-5″-carbaldehyde. The product was purified by column chromatography on silica gel using hexane as the eluent, yielding a red solid (0.42 g; 55% yield).

$^{1}$H-NMR (CDCl$_3$, 300 MHz) δ (ppm) 7,26 (s, 2H), 7.19-7.00 (m, 3H), 7.01-6.81 (m,7H), 6.67 (dd, J=3.4, 1.8Hz, 4H), 2.55-2.39 (m, 12H). $^{13}$C-NMR (CDCl$_3$, 75MHz) δ (ppm) 141.60, 141.57, 140.34, 135.56, 135.07, 134.69, 132.35, 130.76, 127.74, 127.43, 126.80, 126.28, 125.52, 125.37, 124.42, 121.35, 100.01, 15.47, 15.43. HRMS (ESI-TOF; CHCl$_3$/MeOH 10:15); m/z 739.9706 [M+H]$^{+}$ (calcd. for C$_{38}$H$_{28}$S$_8$ 739.9986).

As concerning the film formation of branched thiophene oligomer compounds blended with typical solar cell constituents, several solutions of blended poly(3-hexylthiophene) (P3HT), [6,6]-phenyl C61-butyric acid methylester (PCBM), and BSTV, BOTV were prepared for spin-coating. The preferred solvent was dichlorobenzene (DCB). Each constituent of the blend was first dissolved in a separate tube, and stirred at 50 °C overnight. The solutions were subsequently combined, stirred at 70 °C for 4 h, and overnight at 50 °C. Immediately before the film preparation, the solutions were sonicated for at least 30 min. The "P3HT: PCBM: BSTV (BOTV)" blend was spin-coated in different experimental conditions to optimize the film quality. Thin and reasonably smooth films were obtained when spin-coated at 600 rounds per minute (rpm) for 5 min under nitrogen flow.

The electrochemical measurements, and particularly the differential pulse voltammetry (DPV) experiments, were carried out to calculate the energy levels of the compounds (HOMO and LUMO levels). The basic steady-state spectroscopy (absorption, emission) allowed determination of the main photophysical characteristics of the molecules under investigation, revealing useful indications about usability of these molecules in the photoactive part of the solar cell. This information is essential in designing the photoactive heart of organic solar cells. The DPV measurements of BSTV were carried out in dichloromethane (DCM). BSTV is highly soluble in DCM, which was then used as a solvent also for all the spectroscopic measurements. For BOTV to be soluble in DCM the solution had to be sonicated for at least an hour. Finally, the solutions were filtered, evaporated, and then re-dissolved at lower concentrations.

**RESULTS AND DISCUSSION**

The solar cells for the doping experiments were fabricated and characterized by using standardized and well-known procedures. The device architecture in all the experiments is an organic bulk heterojunction (BHJ) cell with an inverted configuration. An inorganic electron-collecting layer (ZnO) is introduced between the organic BHJ blend and the bottom electrode (Indium Tin Oxide, ITO), causing a drift of the electrons towards the cathode (ITO in this case), and of holes towards the anode (top-electrode metal, Ag in this case). In a "normal configuration" solar cell, holes and electrons are collected at opposite electrodes (ITO and Ag, respectively), and that's why the solar cell architecture used in our experiments is defined *inverted*. Various concentration ratios of BSTV and BOTV as dopants were tested against a reference standard cell made of P3HT: PCBM. From the obtained results, low percentages between 1 and 3% were able to improve the efficiency of the solar cell.

Absorption, emission, and excitation spectra of BSTV and BOTV in DCM solutions are shown in Figure 3. Main absorption/emission bands can be identified, together with their corresponding excitation wavelengths.

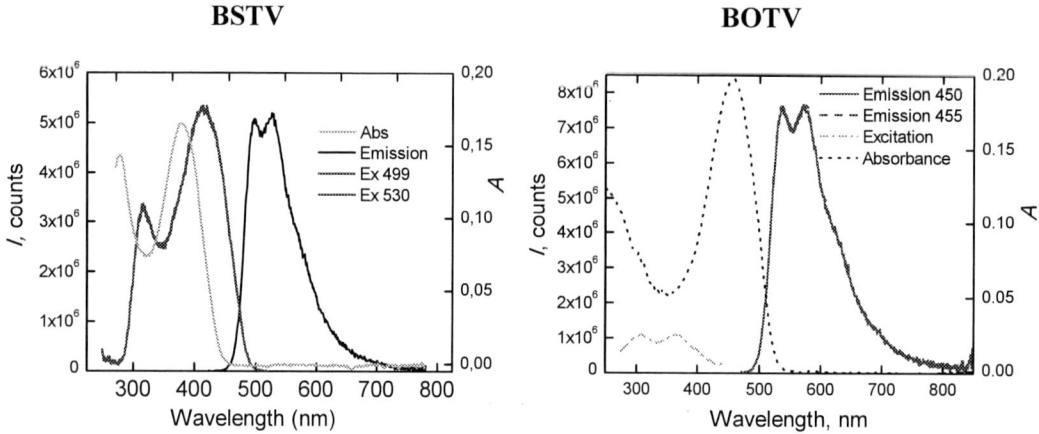

**Figure 3** Absorption, emission, and excitation spectra of BSTV and BOTV in DCM solutions.

The UV-vis absorption maxima of BSTV and BOTV were 427 and 460 nm, respectively, while the emission was close to 550 and 590 nm. respectively. The increasing conjugation length from 6 to 8 thiophene rings induced a red shift of UV-Vis. around 30 nm absorption maximum peak.

The oxidation and reduction curves of the studied compounds, were obtained from Differential Pulse Voltammetry (DPV) experiments (Supporting electrolyte: TBAPF$_6$ 0.1 M, Counter electrode: Graphite, Pseudo-reference electrode: Platinum wire, Working Electrode: Glass Platinum and Ferrocene as internal standard reference). The HOMO and LUMO energy levels were calculated based on the redox potentials referenced to the ferrocene oxidation potential. The results are presented in Table I.

**Table I**. Energy levels (HOMO, LUMO) calculated from the DPV measurements.

| Compound | HOMO (eV) | LUMO (eV) |
|----------|-----------|-----------|
| BSTV | 5.20 | 2.50 |
| BOTV | 5.18 | 3.29 |

When the branched thiophene oligomers compounds are placed in their energetic position among the solar cell components, it is clear that, from an energy level perspective, BSTV and BOTV are more promising because their LUMO levels lie between the PCBM (3.7 eV) and P3HT (2.8 eV) LUMOs. While BSTV has the advantage being easier to use in solar cell preparation being more soluble, the energy of the LUMO is higher than that of P3HT. The high energy level makes it less likely for the efficient electron transfer to occur compared to a solar cell made using BOTV, from a pure energetic perspective.

Different set of experiments were carried out to test the performance of the OPV cells with different amounts of BSTV and BOTV. Several solutions of P3HT: PCBM: BSTV (BOTV) (molar ratios) in dichlorobenzene were prepared with varying concentrations of BSTV or BOTV for use in solar cells.

Tables II and III resume the best results obtained from a set of many experiments with different combinations of the reference cell and the branched thiophene oligomers.

**Table II**. Photovoltaic parameters of solar cells, as derived from J–V curves.

|  | Sample ITO\|ZnO\|BHJ\|Alq$_3$\|Ag | $J_{sc}$ (mA/cm$^2$) | $V_{oc}$ (V) | FF (%) | η (%) |
|--|-------------------------------|----------------------|--------------|--------|-------|
| CH1 | P3HT:PCBM = 6.85:1, reference | -2.91 | 0.58 | 55.56 | 2.76 |
| CH2 | P3HT:PCBM:BSTV = 6.85:1:0.02 | -3.43 | 0.588 | 51.88 | 3.07 |
| CH3 | P3HT:PCBM: BSTV = 6.85:1:0.04 | -2.49 | 0.42 | 31.86 | 0.98 |
| CH4 | P3HT:PCBM: BSTV = 6.85:1:0.06 | -2.57 | 0.588 | 57.72 | 2.57 |

**Table III**. Photovoltaic parameters of solar cells, as derived from J–V curves.

|  | Sample ITO\|ZnO\|BHJ\|Alq$_3$\|Ag | $J_{sc}$ (mA/cm$^2$) | $V_{oc}$ (V) | FF (%) | η (%) |
|--|-------------------------------|----------------------|--------------|--------|-------|
| CH50 | P3HT:PCBM = 6.85:1, reference | -2.78 | 0.54 | 51.58 | 2.28 |
| CH51 | P3HT:PCBM:BOTV= 6.85:1: 0.01 | -3.43 | 0.57 | 58.71 | 3.26 |
| CH52 | P3HT:PCBM:BOTV= 6.85:1: 0.02 | -2.79 | 0.58 | 61.47 | 2.93 |
| CH53 | P3HT:PCBM:BOTV= 6.85:1: 0.03 | -2.49 | 0.58 | 62.07 | 2.65 |

By comparing the results achieved for the OPV cells containing different ratios of BSTV and BOTV with P3HT and PCBM, (Table II-III), it is clear that the best efficiencies ($\eta$) are for experiments CH2, 3.07% (BSTV) and CH51, 3.26% (BOTV) with filled factor (FF) over 50%. In both cases the best results were obtained with the lower amount of the thiophene oligomers. When the content of both branched thiophene oligomers are increased, maintaining P3HT and PCBM content constant, the efficiency ($\eta$) is reduced. That means, over certain amount of the dopant, the photon absorption in no longer advantageous, because the organic material, as is known has a very high absorption coefficient and only thickness of 100 nm are normally necessary for light absorption. Beyond certain amount, the charge mobility plays against. By knowing that charge mobility is very small in organic materials, an optimum ratio of donor (D) and acceptor (A) are required to get the best performance on the OPV cells. When we understand a BHJ material as a solid mixture of two components, D and A, with a nanostructured morphology by spontaneous phase separation, it is necessary to find the best solvent to assure that these D and A components self-assemble to form a continuous interpenetrating network. We think that better performances of the OPV could be achieved with our system if a solvent vapor annealing (SVA) is applied to the devices, as it has been reported recently [11-13] to similar system containing P3HT/PCBM treated with SVA. Here higher performances are reported without damaging the organic components as it could happens with thermal annealing which causes large aggregation of PCBM with a detrimental device performance. [14]

## CONCLUSIONS

This study reports the effect of soluble branched thiophene oligomers, as well defined structures, which can enhance the performance of organic photo cells. It was possible to prepare quite homogeneous films by spin coating and prepare BHJ with P3HT and PCBM, where the performances achieved where close to 3%.

## ACKNOWLEDGMENTS

The authors would like to thank AKA-FONDECYT-ERNC-013 project for financial support.

## REFERENCES

1. Jean-Michel Nunzi, C. R. Physique **3**, 523–542(2002).
2. Yu-Wei Su, Shang-Che Lan, Kung-Hwa Wei, Material Today, Volume 15, Issue 12, 554–562(2012).
3. Omar A. Abdulrazzaq, Viney Saini, Shawn Bourdo, Enkeleda Dervishi and Alexandru S. Biris, Particulate Science and Technology: An International Journal. Volume 31, Issue 5, pp.427-442(2013).
4. Junsheng Yu , Yifan Zheng and Jiang Huang, Polymers **6**, 2473-2509(2014).
5. Venla M., Manninen, Juha P., Heiskanen, Kimmo M., Kaunisto, Osmo E., O. Hormi and Helge J. Lemmetyinen , *RSC Adv.*, ,**4**, 8846-8855(2014).
6. J. C. Bernède, J. Chil. Chem. Soc. **53**, 1549-1564(2008).

7. F. C. Krebs, Sol. Energy Mater. Sol. Cells **93,** 394(2009).

8. P. Kumar and S. Chand, Prog. Photovolt: Res. Appl., **20**, 377(2012).

9. Yu, G., Gao, J., Hummelen, J.C., Wudl, F., Heeger, A.J., Science **270**,1789-1791(1995).

10. Mishra, A., Bäuerle, P., Angew. Chem. Int. Ed. 51, 2020-2067(2012).

11. R. Fitzner, C. Elschner, M. Weil, C. Uhrich, C. Körner, M. Riede, K. Leo, M. Pfeiffer, E. Reinold, E. Mena-Osteritz, P. Bäuerle, Adv. Mater. **24**, 675(2012).

12. Zhang, W ; Hu, R. ; Huo, M.M ;,Ai, X. C.; Zhang, J.P., J. Phys. Chem. C **116**, 4298-4310(2012).

13. Chen, H.; Zang, H.; Hu, B.; Daduum, M.; Adv. Funct. Mater. **23**, 1701-1710(2013).

14. Bertho, S.; Janssen, J.;Cleij, T.J.; Conings, B.; Moons, W.; Gadissa, A.;D´Haen, J.; Goaverts, E.; Lutsen, L.; Manca, J.; Sol.Energ. Mater. Sol. Cells **92**,753-760(2008).

**Mater. Res. Soc. Symp. Proc. Vol. 1737 © 2015 Materials Research Society**
**DOI: 10.1557/opl.2015.540**

# Organic photovoltaic module development with inverted device structure

Shigehiko Mori, Haruhi Oh-oka, Hideyuki Nakao, Takeshi Gotanda, Yoshihiko Nakano, Hyangmi Jung, Atsuko Iida, Rumiko Hayase, Naomi Shida, Mitsunaga Saito, Kenji Todori, Taro Asakura, Akihiro Matsui, and Masahiro Hosoya

Corporate Research & Development Center, Toshiba Corporation, 1, Komukai Toshiba-cho, Saiwai-ku, Kawasaki-shi, 212-8582, Japan

## ABSTRACT

The power conversion efficiency (PCE) of organic photovoltaic (OPV) modules with 9.5% (25 cm$^2$) and 8.7% (802 cm$^2$) have been demonstrated. This PCE of the module exceeded our previous world records of 8.5% (25 cm$^2$) and 6.8% (396 cm$^2$) that were listed in the latest Solar Cell Efficiency Tables ver.43 [1]. Both module design and coating/patterning technique were consistently studied for module development. In order to achieve highly efficient modules, we increased the ratio of photo-active area to designated illumination area to 94% without any scribing process and placed insulating layers in order to decrease the leakage current. The meniscus coating method was used for the fabrication of both buffer and photoactive layers. This technique ensures the fabrication of uniform and nanometer order thickness layers with thickness variation less than 3%. Furthermore, the PCE of the OPV under indoor illumination was found to be higher than that of the conventional Si type solar cells. This indicates that OPVs are promising as electrical power supplies for indoor applications. Therefore, we have also developed several prototypes for electronics integrated photovoltaics (EIPV) such as electrical shelf labels and wireless sensors embedded with our OPV modules, which can be operated by indoor lights.

## INTRODUCTION

Organic photovoltaics (OPVs) are one of the most promising candidates for next generation solar cells because they are light, flexible, and therefore able to be fabricated by roll-to-roll processes showing low-cost. Although OSC has been suffered from the low power conversion efficiency (PCE), recently the PCE exceeding 10% has been reported [1-3]. Therefore, momentum of commercial realization for OPV has been growing in these days. Note that the development of the OPV module with large active area is indispensable for commercialization, but on the other hand, there have been little studies of OPV module. In this paper, we report on the development of OPV modules whose active area was ranging from 25 cm$^2$ to 802 cm$^2$ with high performance. Additionally, OPV showed a remarkable advantage that electric power generation performance under indoor light illumination is superior to that of the conventional solar cells [4]. We also show several indoor applications of OPV modules for electrical power supplies.

## EXPERIMENT AND DISCUSSION

### Cell development

For the development of module, we firstly investigated the OPV cell structure whose active area was 1 cm$^2$. Our sample was composed of anti-reflection film/glass/ITO/hole transport layer/positive type polymer : PC$_{70}$BM BHJ/electron transport layer/metal layer (PC$_{70}$BM:[6,6]-phenyl-C71-butyric acid methyl ester). Our positive type polymer is a donor-acceptor type polymer synthesized by our group having both deeper highest occupied molecular orbital (HOMO) energy level and small energy band-gap.

The electrical properties of the OPV were measured under 1 sun (AM-1.5) and indoor illumination. Here, a daylight color light-emitting diode (LED) was used as an indoor light source. The 1 sun spectrum ranged from ultraviolet to infrared optical wavelengths, while the daylight color LED spectrum ranged from 400 to 800 nm, which was much narrower than 1 sun. Also, current-voltage (J-V) characteristics of crystal silicon solar photovoltaics (c-Si-PV) and amorphous silicon photovoltaics (a-Si-PV) were measured for comparison. Each light source was coupled into a fiber. The emerging light from the fiber which passed through a lens was incident to the cell from the anti-reflection film side. Here, a mask which defines the active area ($10 \times 10$ mm$^2$) was attached to the anti-reflection film side for accurate measurement. 1 sun and LED intensity was determined by using a crystal silicon reference cell (Konika minolta, AK-200) and a spectroradiometer (Eko instruments, MS-720), respectively. Then, the J-V curves were recorded during light irradiation. In our experiments, we used the AK-200 and AM-1454CA (Panasonic) as a representative c-Si-PV and a-Si-PV in order to compare the J-V characteristic with OPV, respectively. Because AM-1454 is composed of four serial cell connections, we measured the J-V response of only its single cell. Table 1 shows the J-V characteristics of c-Si-PV, a-Si-PV, and OPV when illuminated by 1 sun or LED.

Table 1. Measured PCE, J$_{sc}$, V$_{oc}$, and FF of c-Si-PV, a-Si-PV, and OPV with various incident light intensities. P$_{in}$ and P$_{out}$ indicate the incoming power of the light and output power of the solar cell, respectively.

| Solar cell type | c-Si-PV | c-Si-PV | a-Si-PV | a-Si-PV | OPV | OPV |
|---|---|---|---|---|---|---|
| Light source intensity | 1 sun | 209 lx | 1 sun | 209 lx | 1 sun | 223 lx |
| PCE (%) | 14.10 | 9.19 | 0.43 | 10.51 | 11.54 | 14.26 |
| J$_{sc}$ (mA/cm$^2$) | 31.80 | 0.02 | 2.15 | 0.02 | 19.83 | 0.021 |
| V$_{oc}$ (V) | 0.60 | 0.37 | 0.86 | 0.61 | 0.79 | 0.60 |
| FF | 0.74 | 0.64 | 0.23 | 0.66 | 0.74 | 0.75 |
| P$_{in}$ (W/cm$^2$) | 0.1 | $6.29 \times 10^{-5}$ | 0.1 | $6.29 \times 10^{-5}$ | 0.1 | $6.67 \times 10^{-5}$ |
| P$_{out}$ (W/cm$^2$) | $1.41 \times 10^{-2}$ | $5.78 \times 10^{-6}$ | $4.25 \times 10^{-4}$ | $6.60 \times 10^{-6}$ | $1.15 \times 10^{-2}$ | $9.52 \times 10^{-6}$ |

In the case of 1 sun, we observed that the PCE of c-Si-PV, a-Si-PV, and OPV were 14.1%, 0.4%, and 11.5%, respectively. Because a-Si-PV is recommended for indoor use, a low PCE could be observed. As for OPV, owing to use the polymer possessing on deeper HOMO energy level and small energy band-gap, purified PC$_{70}$BM, and band matched metal oxide buffer layers, we obtained a high performance cell with PCE of 11.5%. When a LED light was used, the OPV exhibited the highest PCE among them. Because the c-Si-PV had wider wavelength coverage than the a-Si-PV and OPV, it exhibited the largest value of J$_{sc}$ at 1 sun. In contrast, the emission spectrum of the LED ranged from 400 to 800 nm and this was within the EQE curve range of both a-Si-PV and OPV. As a result, the J$_{sc}$ difference among three types of solar cell became much smaller under LED illumination. As can be seen in Table 1, V$_{oc}$ decreased with decreasing light intensity. It is known that V$_{oc}$ is proportional to the logarithm of incoming light intensity. Because the energy band gap of a-Si-PV is larger than that of c-Si-PV, a-Si-PV exhibited a higher $V_{oc}$. Here, the $V_{oc}$ of OPVs is also proportional to the difference between the HOMO energy level of a positive type organic semiconductor polymer and the lowest unoccupied molecular orbital (LUMO) energy level of a negative type organic semiconductor material. In our experiments, the higher $V_{oc}$ values of the OPV were observed because of the deep HOMO energy level of polymer we used (~5.3 eV). To summarize, due to the high values of $J_{sc}$ and $V_{oc}$ under LED illumination, the OPV exhibited superior properties. Furthermore, let us note that OPV showed high PCEs under both 1 sun and indoor light illuminance.

### Module development

For module development, we have investigated the module structure design and the coating/patterning method. Module structure design is important for achieving a high FF value, while coating and patterning technique is also indispensable for uniform coating and high-density patterning without scribing.

In module design, we first fabricated the OPV module normally. However, the efficiency was low mainly because of the low fill factors. Therefore, we attempted to develop a module structure enabling the demonstration of a high FF value. In OPV module, the striped cells are connected in series in the direction perpendicular to the stripes and the cathode is connected to the anode of the next striped cell. If there exist short circuits or disconnections between electrodes, the fill factor will decrease. Therefore, we placed insulating and conductive layers at appropriate positions. As a result, the FF was improved. However, the active area ratio in the total area was still low in 72%.

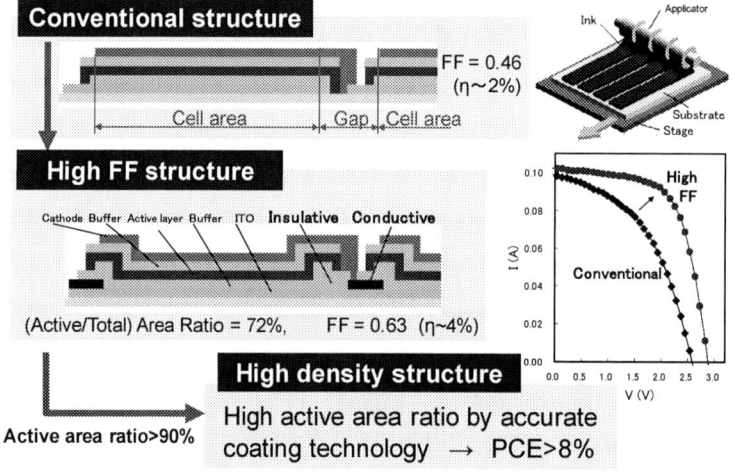

Figure 1: Schematic of module design and meniscus coating.

Next, we developed a high density structure so that the active area ratio exceeds 90%. In the sight of practical process, coating and patterning technology is very important to establish the method of fabricating thin and uniform layers. We developed the "meniscus coating method" shown in Fig. 1. Using this method, the active area ratio up to 94% is obtained, because the striped cells with high width accuracy and low thickness variation can be performed. Moreover, any scribing was not necessary in this process and consumption of solutions for buffer and photo active layers was reduced to about two orders of magnitude. Thus, this method is able to provide a simple and inexpensive manufacturing process.

Using these technologies, we firstly developed the module prototype of 20 cm square with active area ratio of 90%. The module PCE was 6.8% in the official measurement by the national institute of Advanced Industrial Science and Technology (AIST). This value was recognized as the world record since the publication of Solar Efficiency Tables ver. 41. Next, we developed 5 cm square sub-modules with the active area ratio of 94% (Fig. 2(a)). As a result, the efficiency went up to 9.5% certificated by AIST. Furthermore, we are developing 30 cm square modules, aiming at PCE over 10%, for practical applications. Recently, PCE of 8.7% officially measured by AIST has been demonstrated (see Fig. 2(b)).

(a)                                                          (b)

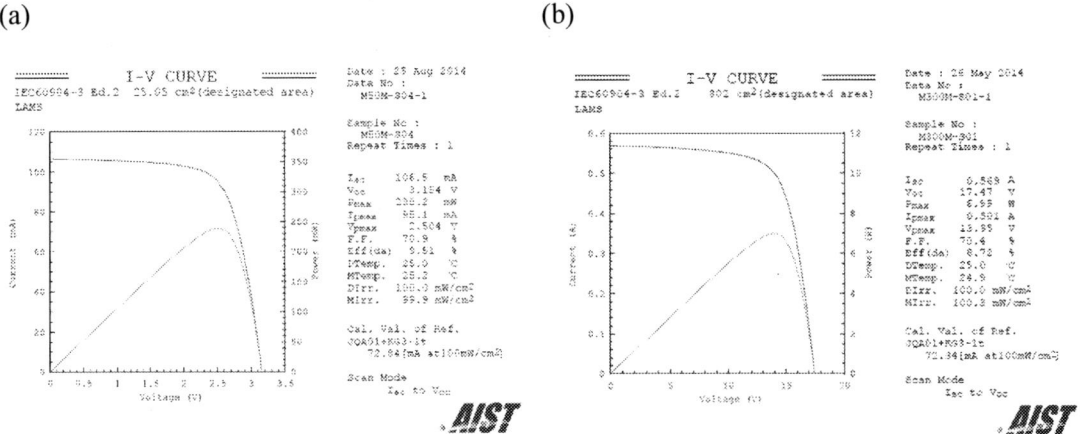

Figure 2: Measured results of (a) 5 cm square sub-module and (b) 30 cm square module.

Figure 3 shows the PCE changes of world records listed in Solar Cell Efficiency Tables and those measured in Toshiba. The unfilled and filled blue square points denote the best measured OPV cell and sub-module PCEs, respectively. These data were taken from the Solar Cell Efficiency Tables. The unfilled and filled red circle points denote the best measured cell and sub-module PCEs in our laboratory, respectively. In the curve of OPV module, Toshiba's efficiencies of 6.8% and 8.5% were recognized as the latest records. Furthermore, as mentioned above, the latest PCEs of OPV modules developed in our group have reached 9.5% (25 cm$^2$) and 8.7% (802 cm$^2$).

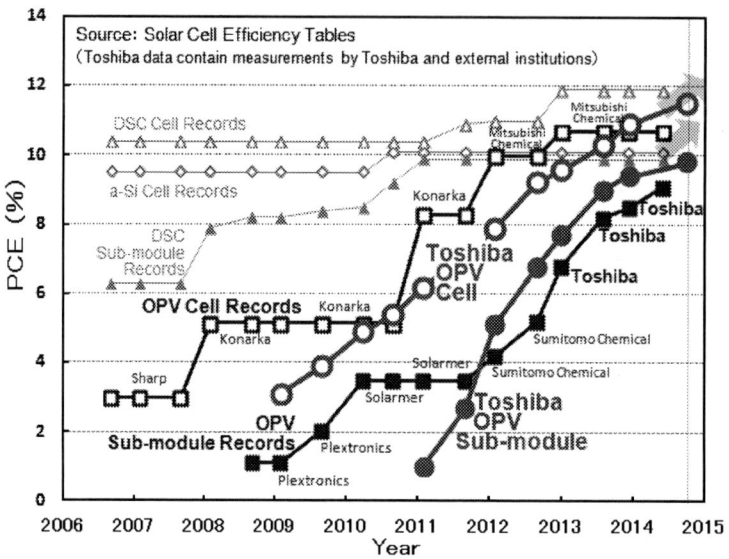

Figure 3: Best research photovoltaics efficiencies from 2006 to present. The unfilled and filled red square points denote the best measured cell and sub-module PCEs in our laboratory, respectively. The other data were taken from the Solar Cell Efficiency Tables.

**Durability tests**

For practical use of OPV, we are performing the durability tests based on JIS-8938 (Japanese Industrial Standards). Five types of tests, namely 1. thermal cycling test, 2. temperature/humidity cycling test, 3. light soaking test, 4. heat test, and 5. damp/heat test have been demonstrating. In order to pass these tests, relative reduction of PCEs less than 10% must be achieved. It is said that passing these tests are equivalent to 10-year durability test. We had obtained good results in 1 cm$^2$ cell, and now we are carrying out these tests in sub-module.

**Applications of OPV**

As denoted in Table 1, OPV cell showed superior performance when an indoor light was used. Evaluation of the J-V characteristics for OPV modules under LED operation was performed and similar properties as 1 cm$^2$ cell showed were confirmed. Taking notice that OPV exhibited good features under both outdoor and indoor operation, we are developing several applications of OPV. Among the various applications shown in Fig. 4 (a), indoor based Electronics integrated photovoltaics (EIPV) is expected as one of the promising candidates.

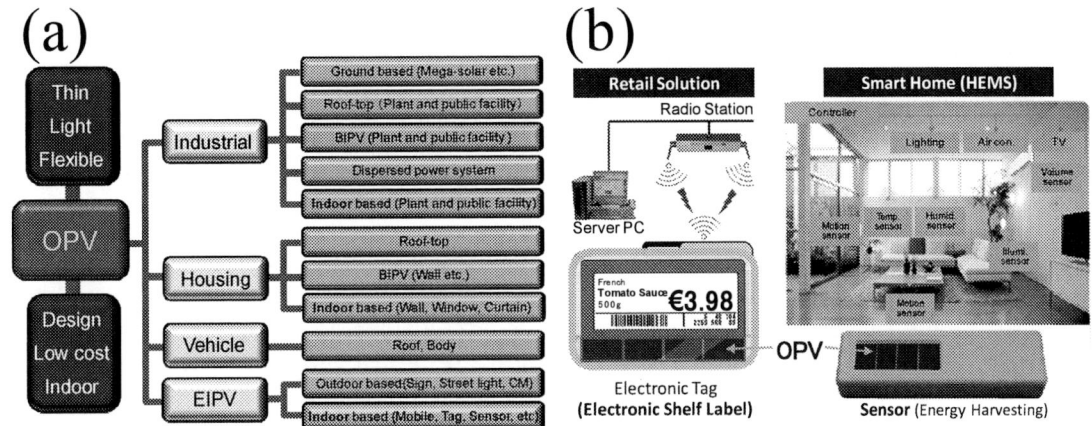

Figure 4: (a) Applications of OPV. (b) Prototypes of EIPV using OPV

We have developed some prototypes for EIPV. The prototypes of electronic shelf labels and sensors were fabricated using our OPV (Fig. 4(b)). It was confirmed that these devices can be driven by power generation of OPV under room lighting. It is expected that these devices will be adopted for practical use in the near future.

## CONCLUSIONS

We have developed OPV minimodule and sub-module exhibiting PCEs of 9.5% (25 cm$^2$) and 8.7% (802 cm$^2$), respectively. For OPV cell (1 cm$^2$), its PCE reached to 11.5%. In indoor operation, OPV showed better performance compared to c-Si-PV and a-Si-PV. Additionally, let us note that OPV is able to generate electrical power efficiently in a wide range of incoming light power (intense 1 sun light to weak indoor light). Application developments and durability tests are also in progress. It is expected that OPV modules of high quality and low cost will be commercially available in the near future.

## ACKNOWLEDGMENTS

This work was supported by the New Energy and Industrial Technology Development Organization (NEDO) project of "Solar power system next generation high-performance technology development".

## REFERENCES

1. M. A. Green, K. Emery, Y. Hishikawa, W. Warta, and E. D. Dunlop, Solar cell efficiency tables (version 43), Prog. Photovolt: Res. Appl. **22**, 1 (2014).
2. M. Hosoya, H. Oooka, H. Nakao, S. Mori, T. Gotanda, N. Shida, R. Hayase, Y. Nakano and M. Saito, Proceedings of 28th European Photovoltaic Solar Energy Conference, 2013, pp. 2236-2238.
3. C.-C. Chen, W.-H. Chang, K. Yoshimura, K. Ohya, J. You, J. Gao, Z. Hong and Y. Yang, Adv. Mater. **26**, 5670 (2014).
4. R. Steim, T. Ameri, P. Schilinsky, C. Waldauf, G. Dennler, M. Scharber, C. J. Brabec, Sol. Energy Mater. Sol. Cells **95**, 3256 (2011).

**Mater. Res. Soc. Symp. Proc. Vol. 1737 © 2015 Materials Research Society**
**DOI: 10.1557/opl.2015.561**

## Effective Mg:Ag / MoO₃ recombination zone for tandem organic photovoltaic devices

A. R. Jeong, S. Wiesner, S. Fengler, M. Ch. Lux-Steiner, M. Rusu[*]
Institut für Heterogene Materialsysteme, Helmholtz-Zentrum Berlin für Materialien und Energie,
Lise-Meitner Campus, Hahn-Meitner-Platz 1, 14109 Berlin, Germany

### ABSTRACT

We demonstrate an effective recombination zone consisting of Mg:Ag (1:3) alloy and MoO₃ layers with 0.8 nm and 3 nm respectively for application in tandem organic photovoltaic devices based on zinc phthalocyanine (ZnPc) donor and fullerene $C_{60}$ acceptor. The Mg:Ag layer ensures an optimum electron selectivity, while MoO₃ layer effectively selects holes. A conversion efficiency of 2.2% has been achieved under an illumination of 100 mW/cm² at room temperature. The open circuit voltage of 810 mV is close to the sum of the open circuit voltages of the constituent single cells. The recombination Mg:Ag-MoO₃ layer system is investigated with regard to the requirements of high optical transparency, work function compatibility, and facilitation of light absorption. The respective characterizations were carried out by UV-Visible spectroscopy, Kelvin probe force microscopy in ultrahigh vacuum, current-voltage and external quantum efficiency methods.

### INTRODUCTION

Organic photovoltaics (OPV) have attracted much attention due to potential light-weight, low cost, transparency and flexible alternative to conventional inorganic solar cells. However, the absorber thickness in single solar cells is constrained by the exciton diffusion length which is much smaller than the light absorption length [1,2]. Therefore, maximum theoretical efficiencies are difficult to achieve on single photovoltaic devices. Tandem solar cells, which consist of two or more single cells with complimentary absorptions, have been considered as an effective approach to overcome this limitation. For the configuration of tandem solar cells, series connected architectures have been widely used. Tandem devices have already achieved high efficiencies of 12%, which means that tandem organic photovoltaic cells are promising candidates for commercial applications [3]. In these architectures, utilization of effective recombination zones for the realization of the series connection between the sub-cells is one of the key aspects to achieve high-efficient tandem devices.

In this work, we demonstrate an efficient recombination zone with ultra-thin Mg:Ag (1:3) and MoO₃, which was reported as electron transport [4,5] and hole transport layers [6-9], respectively. The recombination zone indicates high transparency, proper band alignment between front and back cells, and adequate photoresponse within zinc phthalocyanine (ZnPc) and $C_{60}$ regions.

### EXPERIMENT

The OPV devices were prepared with ZnPc and $C_{60}$ by organic vapor phase deposition (OVPD) [10]. The organic layers were deposited on cleaned indium tin oxide (ITO) substrates coated by poly (3,4-ethylenedioxythiophene):poly (styrenesulfonate) (PEDOT:PSS) layers prepared by spin-coating. The recombination zone was prepared by sequentially depositing Mg:Ag and $MoO_3$ thin films by thermal evaporation in vacuum at about $10^{-6}$ mbar. The final structure of the tandem OPV device was ITO / PEDOT:PSS / ZnPc:$C_{60}$ (60 nm) / $C_{60}$ (10 nm) / Mg:Ag and $MoO_3$ (0.8 and 3 nm) / ZnPc (5 nm) / ZnPc:$C_{60}$ (80 nm) / Mg:Ag (100 nm). The front and back reference cells had the structure of ITO / PEDOT:PSS / ZnPc:$C_{60}$ (60 nm) / $C_{60}$ (10 nm) / Mg:Ag (100 nm) and ITO / PEDOT:PSS / ZnPc (5 nm) / ZnPc:$C_{60}$ (80 nm) / Mg:Ag (100 nm), respectively.

The current-voltage characteristics and external quantum efficiency (EQE) were measured with a Halogen lamp in a nitrogen filled glove box. The measurements of the transparency of the recombination zone were carried out with UV-Visible spectroscopy (Perkin-Elmer 750). The morphology and work function of samples were simultaneously investigated by Kelvin probe force microscopy (KPFM) in ultrahigh vacuum (UHV). Contact potential difference (CPD) between sample and tip was measured and the work function of the sample $\Phi_{sample}$ was determined with the known work function of the tip $\Phi_{tip}$ according to the following equation:

$$\Phi_{sample} = \Phi_{tip} + e V_{CPD,} \tag{1}$$

where $e$ is the elementary charge [11]. The work function of the tip was determined by calibration measurements against a highly oriented pyrolytic graphite (HOPG) sample. For KPFM measurements, the samples were prepared during the same process for OPV devices and sealed inside a glove box for the transport. The samples were loaded into the KPFM chamber with ultra-high vacuum of $\approx 10^{-10}$ mbar without any exposure to the air.

## DISCUSSION

The band diagram of the tandem device is illustrated in Figure 1, where the device structure can be recognized. The values for positions of highest occupied molecular orbitals (HOMO) and lowest occupied molecular orbitals (LUMO) were taken from Refs. 12 and 13. In the front sub-cell, excitons are dissociated into free carriers at the ZnPc:$C_{60}$ interface. The generated holes are transported through ZnPc to the PEDOT:PSS / ITO interface, while electrons are transported through $C_{60}$ to the Mg:Ag layer. In the back cell, exitons are dissociated in the mixed ZnPc:$C_{60}$. Electrons are transported to the $C_{60}$ / Mg:Ag (1:3) interface, where $MgAg_3$ has been reported to form a good Ohmic contact to $C_{60}$ and a blocking contact to ZnPc as well as proper work function as a cathode material [5]. On the other hand, holes are transported to $MoO_3$ / ZnPc interface, where they recombine with electrons collected on Mg:Ag layer from the front sub-cell. Here, the Fermi levels of Mg:Ag and $MoO_3$ are aligned resulting in the band alignment shown in Figure 1.

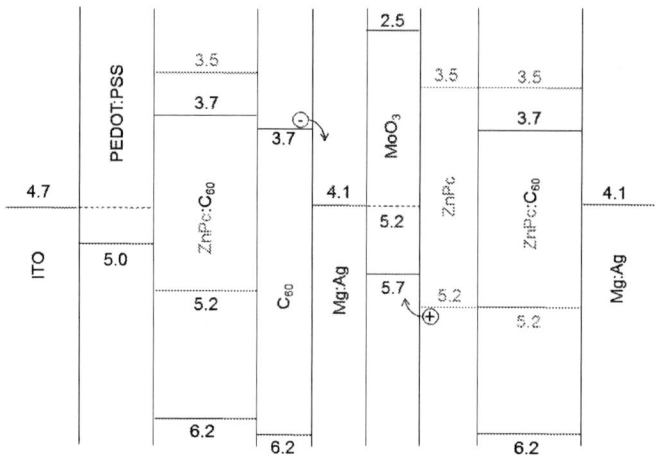

**Figure 1 Energy level diagram of the tandem organic photovoltaic device with Mg:Ag-MoO₃ intermediate layer.**

Figure 2 shows the measured current-voltage characteristics of tandem OPV (open square) measured under 100 mW/cm². It shows a $V_{oc}$ = 810 mV, FF = 39%, $J_{sc}$ = 6.88 mA/cm² and Eff = 2.2%, where $V_{oc}$ is open circuit voltage, FF is fill factor, $J_{sc}$ is short-circuit current density and Eff is power conversion efficiency. The $V_{oc}$ of the tandem device is almost summation of the $V_{oc}$-s of both sub-cells, which shows that the charge recombination process in created Mg:Ag-MoO₃ recombination zone occurs efficiently. The Mg:Ag-MoO₃ system has high transmittance of about 90% in the spectral range from 300 to 800 nm (not shown). This demonstrates that the optimized recombination zone is suitable as intermediate layer for tandem applications. For the front sub-cell with the thickness of 60 nm ZnPc:C₆₀ and 10 nm C₆₀ (close circle), an efficiency of 1.5% is achieved with the $V_{oc}$ = 430 mV, $J_{sc}$ = 9.66 mA/cm², and FF = 36%. The back cell with 80 nm ZnPc:C₆₀ (close triangle) records the Eff = 1.8%, $V_{oc}$ = 440 mV, $J_{sc}$ = 10.26 mA/cm², and FF = 40%. Low shunt resistance of the front reference cell could be influenced by the PEDOT:PSS / ZnPc:C₆₀ interface, while the back reference cell has a different interface, namely PEDOT:PSS / ZnPc. In the letter stack, ZnPc layer can function as a buffer layer to reduce the leakage current. The compatible current density of the front and back cells demonstrates that thicknesses of the absorber layers are adequate. The total current of the tandem device is expected in this case to be approximately the same. However, the $J_{sc}$ of the tandem solar cell is by about 30% reduced. This decrease of $J_{sc}$ in the tandem cell is explained by losses related to the absorbed back-reflected light from the Mg:Ag rear electrode. Further optimization of the absorber and buffer layers should be performed for the better match of the absorber position to the optical field distribution across the device. Because we have focused in this work on the investigation of the recombination zone, similar absorber materials were chosen for the front and back cells. For tandem devices consisting of different absorber materials with absorption spectra matched to different regions of the visible spectrum, the same components MoO₃ and Mg:Ag of the recombination zone could be proposed: due to its high work function of 5.2 eV, MoO₃ with $E_c$ = 2.5 eV and $E_v$ = 5.7 eV [9] could be applied as hole collecting layer for a large variety of donor organic materials; the work function of the Mg:Ag alloy can be easily tuned [4] to fit electronically as electron collecting layer do different acceptor organic materials.

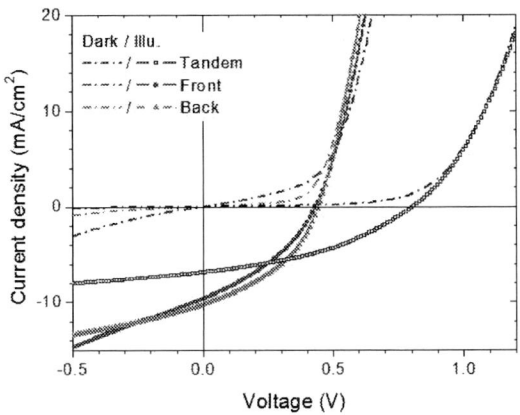

**Figure 2 Current-voltage characteristics under 100 mW/cm$^2$ for the front-only (circle), back-only (triangle), and tandem (square) organic photovoltaic cells.**

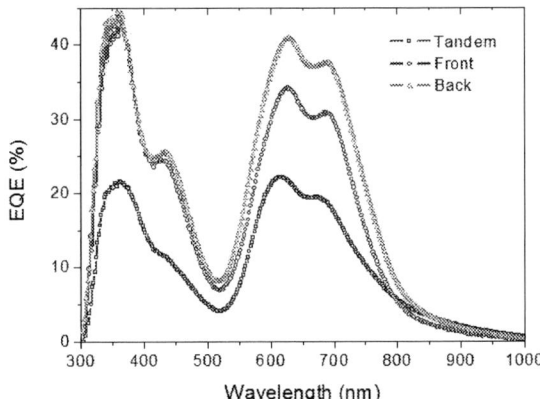

**Figure 3 EQE spectra for the front-only (circle), back-only (triangle), and tandem (square) organic photovoltaic cells.**

The experimental external quantum efficiency (EQE) spectra of the tandem cell with an optimized recombination zone and of the corresponding reference cells are shown in Figure 3. Photoresponse of the back cell in the region over 550 nm is higher than that of the front cell because an additional ZnPc layer is added in the back cell. Meanwhile, photoresponse region of $C_{60}$ below 550 nm shows comparable EQE spectra of the front and back sub-cells. The transport of electrons from the front cell with 60 nm ZnPc:$C_{60}$ / 10 nm $C_{60}$ and from the back cell with 80 nm ZnPc:$C_{60}$ could be equivalent. On the other hand, the EQE spectrum of the tandem device is nearly half of the reference single cells because of the reasons discussed above for differences observed for $J_{sc}$.

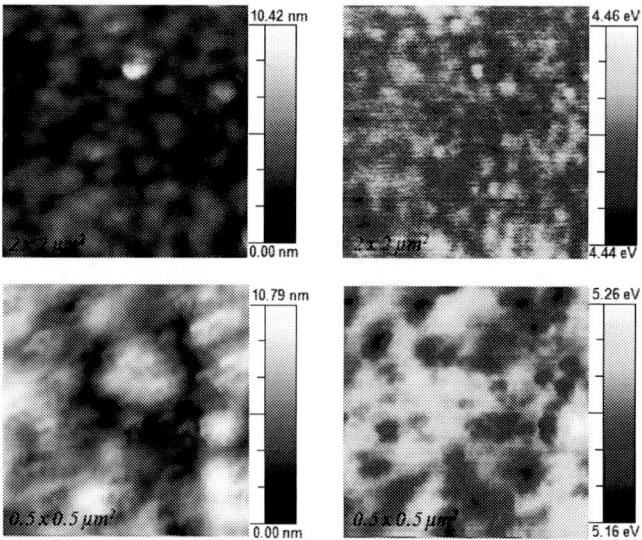

**Figure 4 Topography and work function of the front cell of the tandem device for (a), (b) as-prepared ITO / PEDOT:PSS / 60 nm ZnPc:C$_{60}$ / 10 nm C$_{60}$ stack and (c), (d) annealed at 150°C for 30 minutes ITO / PEDOT:PSS / 60 nm ZnPc:C$_{60}$ / 10 nm C$_{60}$ / Mg:Ag-MoO$_3$ stack.**

To investigate the electronic properties of the recombination zone, its work function and that of sub-layers was measured by KPFM. The work function of the active layer in the front cell was investigated on an as-prepared thin film. The work function of the recombination zone was measured after heating the ITO / PEDOT:PSS / 60 nm ZnPc:C$_{60}$ / 10 nm C$_{60}$ / Mg:Ag-MoO$_3$ stack at 150°C for 30 minutes to simulate the conditions to what the front cell was exposed during the preparation of the back cell. In Figure 4 (a) and (b), the work function of the as-prepared blend organic layer is determined of 4.450 ± 0.005 eV, which correlates with the literature data for ZnPc:C$_{60}$ [7]. It also exhibits very smooth surface with root mean square (RMS) roughness of 1.0 nm. On the other hand, the work function of the top surface of the intermediate layer on the organic surface significantly increases to 5.21 ± 0.01 eV with higher RMS roughness of 1.9 nm after heating in the ultra-high vacuum. The latter work function corresponds to values we have recorded on thick (≥10 nm) MoO$_3$ thin films (not shown). The corresponding topography and work function images are shown in Figure 4 (c) and (d). The work function of the Mg:Ag (1:3) layer of 4.1 eV was measured in our previous work [4, 5], which is suitable for an electron transport layer due to lower value than the work function of C$_{60}$. On the other hand, MoO$_3$ has been well known as a hole transport layer [6-9]. In addition, according to Ref. 9, the E$_c$ and E$_v$ of MoO$_3$ are at 2.5 eV and 5.7 eV and therefore, the ultra-thin MoO$_3$ layer can block the electrons but collect the holes. Thus, in the intermediate layer the electrons collected on Mg:Ag will recombine with the holes collected on MoO$_3$. Therefore it is clear that the recombination zone consisting of Mg:Ag and MoO$_3$ layer can be an efficient intermediate layer for organic photovoltaic applications.

**CONCLUSIONS**

We have demonstrated tandem OPV devices with an efficient Mg:Ag-MoO$_3$ layer system, which realizes the recombination of electrons from the front cell with the holes from the back cell. The ultra-thin Mg:Ag-MoO$_3$ layer leads to a proper band alignment between the front and back cells. The resulting tandem device demonstrated Eff = 2.2% with an open circuit voltage of 810 mV, which is close to the sum of the V$_{oc}$-s of the front and back sub-cells. Since the electronic properties of the Mg:Ag alloy can be tuned in a wide range to fit the requirements for an electron collecting layer from acceptor materials and MoO$_3$ optical and electronic properties facilitate collection of holes from a wide range of donor materials, the Mg:Ag-MoO$_3$ recombination system can be suggested for the application in tandem cells based on a large variety of organic materials.

## ACKNOWLEDGMENTS

This work has been supported by the EU under the NMP Project Smartonics No. 310229.

## REFERENCES

1. P. Peumans, S. Uchida, and S. R. Forrest, *Nature (London)* **158**, 425 (2003).
2. P. Peumans, A. Yakimov, and S. R. Forrest, *J. Appl. Phys.* **93**, 3693 (2003).
3. Heliatek. Available online: http://www.heliatek.com/
4. M. Rusu, S. Wiesner, I. Lauermann, Ch.-H. Fischer, K. Fostiropoulos, J. N. Audinot, Y. Fleming, and M. Ch. Lux-Steiner, *Appl. Phys. Lett.* **97**, 073504 (2010).
5. M. Rusu, F. Kraffert, S. Wiesner, W. Schindler, K. Fostiropoulos, M. Ch. Lux-Steiner, *Energy Procedia* **31**, 96 (2012).
6. J. Meyer , S. Hamwi, M. Kröger, W. Kowalsky, T. Riedl, and A. Kahn, *Adv. Mater.* **24**, 5408 (2012).
7. M. Pfeiffer, K. Leo, and N. Karl, *J. Appl. Phys.* **80**, 6880 (1996).
8. Y. Sun, C. J. Takacs, S. R. Cowan, J. H. Seo, X. Gong, A. Roy, and A. J. Heeger, *Adv. Mater.* **23**, 2226 (2011).
9. J. K. Larsen, H. Simchi, P. Xin, K. Kim, and W. N. Shafarman, *Appl. Phys. Lett.* **104**, 033901 (2014).
10. M. Rusu, S. Wiesner, T. Nete, H. Blei, N. Meyer, M. Heuken, M. C. Lux-Steiner, and K. Fostiropiulos, *Renewable Energy* **33**, 254 (2008).
11. S Sadewasser and T. Glatzel, *Kelvin Probe Force Microscopy Measuring and Compensating Electrostatic Forces*, (Springer Series in Surface Sciences, Heidelberg (2012)) p.12-14.
12. K. L. Mutolo, E. I. Mayo, B. P. Rand, S. R. Forrest, and M. E. Thompson, *J. Am. Chem. Soc.* **128**, 8108 (2006).
13. Z. R. Hong, R. Lessmann, Maennig, Q. Huang, K. Harada, M. Riede, and K. Leo, *J. Appl. Phys.* **106**, 064511 (2009).

# AUTHOR INDEX

Asakura, T. .................................................. 26

Autreto, P. A. S. ......................................... 13

Beingessner, R. L. ........................................ 1

Cho, J.-Y. ...................................................... 1

Efimov, A. ................................................... 19

Fengler, S. .................................................... 32

Fenniri, H. ..................................................... 1

Freire, J. A. ................................................. 13

Galvao, D. S. .............................................. 13

Gotanda, T. ................................................. 26

Hayase, R. .................................................... 26

Hosoya, M. ................................................... 26

Iida, A. .......................................................... 26

Jeong, A. R. ................................................. 32

Johnson, B. .................................................... 7

Jung, H. ......................................................... 26

Lemmetyinen, H. ......................................... 19

Lux-Steiner, M. Ch. .................................... 32

Machado, L. D. .......................................... 13

Martinez, F. ................................................ 19

Matsui, A. ..................................................... 26

Mori, S. ......................................................... 26

Nakano, Y. ................................................... 26

Nakao, H. ..................................................... 26

Neculqueo, G. ............................................. 19

Oh-Oka, H. .................................................. 26

Ostroverkhova, O. ........................................ 7

Parthasarathy, V. .......................................... 1

Paudel, K. ...................................................... 7

Rusu, M. ....................................................... 32

Saito, M. ....................................................... 26

Shida, N. ....................................................... 26

Shuai, L. ......................................................... 1

Todori, K. ..................................................... 26

Vasquez, S. O. ............................................. 19

Vivo, P. ......................................................... 19

Wiesner, S. ................................................... 32

Woellner, C. F. ........................................... 13

Yamazaki, T. .................................................. 1

Cambridge University Press
32 Avenue of the Americas, New York, NY 10013-2473, USA

Materials Research Society
506 Keystone Drive, Warrendale, PA 15086

ISBN 978-1-5108-0617-7